MANAGING MANURE

How to Store, Compost, and Use Organic Livestock Wastes

Mark Kopecky

Storey Publishing

The mission of Storey Publishing is to serve our customers by publishing practical information that encourages personal independence in harmony with the environment.

Edited by Deborah Burns
Series design by Alethea Morrison
Art direction by Jeff Stiefel
Text production by Theresa Wiscovitch
Indexed by Northwind Editorial Services

Cover illustration by © Lisel Ashlock
Interior illustrations by © Steve Sanford

Storey Publishing
210 MASS MoCA Way
North Adams, MA 01247
www.storey.com

Printed in the United States by McNaughton & Gunn, Inc.
10 9 8 7 6 5 4 3 2 1

Library of Congress Cataloging-in-Publication Data

Kopecky, Mark, author.
Managing manure : how to store, compost, and use organic livestock wastes / by Mark Kopecky.
pages cm. — (A Storey basics title)
Other title: How to store, compost, and use organic livestock wastes
Includes index.
ISBN 978-1-61212-458-2 (pbk. : alk. paper)
ISBN 978-1-61212-459-9 (ebook) 1. Farm manure. 2. Animal waste. I. Title. II. Title: How to store, compost, and use organic livestock wastes. III. Series: Storey basics.
S655.K66 2015
631.8'61—dc23

2014033683

CONTENTS

For Howard:
Teacher, mentor, brother, friend.

INTRODUCTION

BROWN GOLD

Manure is part of agriculture, and part of the incredible web of life. Although society considers it a waste product, most farmers and gardeners view manure as a priceless resource.

Like all components of an ecological system, it's impossible to deal with the subject of manure in isolation. In this book, I'll try to make some bridges between manure management and crop production, food quality, and human and environmental health. This isn't an exhaustive reference work, but I hope it will give small farmers and gardeners practical information to help you appreciate and make the most of this wonderful resource.

One note of caution: Throughout this book I'll make references to average values for various characteristics of manure. The values I refer to here are from reputable sources, mostly various universities, and are often the result of thousands of measurements. It's good to remember, though, that manure is notoriously variable, and the characteristics of the version you have to work with might be very similar to these averages or quite different. To find out more about the manure on your farm, you can have it tested (I'll talk more about that in chapter 3).

MANURE HAPPENS

WHEREVER LIVING THINGS GO about their daily business, manure happens. Physics teaches us that in every ordinary system, matter is neither created nor destroyed. A corollary of this rule is that every living thing produces some by-product of its existence that needs to be accounted for and somehow reused. Broadly speaking, even plants produce "waste" products (oxygen), which illustrates how essential one organism's waste can be for the very existence of others. Even tiny creatures like nematodes, protozoa, fungi, and bacteria excrete waste products that are very important and beneficial for the life around them.

What Is Manure?

For the purposes of this book, I'll use the definition of manure that most people relate to: the dung (feces) and urine, mostly of farm animals, with or without other materials used as bedding. (There is also green manure, which is different: cover crops planted and turned under specifically to fertilize the soil.)

Manure can be a source of fresh organic matter and nutrients that benefit the soil ecosystem and its crop component. It can also contain pathogens that can harm either the plants that grow in the soil or the livestock or people who eat them. Like most components of biological systems, manure has many beneficial traits along with a few we need to be careful of.

MANURE IS A COMBINATION of digested, partially digested, and undigested remains of the food that an animal eats, along with a broad mix of microorganisms.

An Age-Old Soil Amendment

For thousands of years, manure has been an integral part of agriculture. It is mentioned a number of times in the Bible. Many of the passages refer to human excrement and serve as admonitions or slurs against unholy practices, but at least two examples from the book of Luke refer to manure as a soil **amendment**. In the 1st century AD, a Roman soldier-turned-farmer and historian, Lucius Junius Moderatus Columella, wrote extensively in 12 books about agricultural practices in *De re rustica*. The University of Wisconsin Extension publication *Management of Wisconsin Soils* provides the following quote from Columella:

> And by plentiful dunging, which is owing to flocks and herds of cattle, the earth produces her fruit in great abundance.

That statement captures the value and benefits of manure beautifully.

Going back even further, traditional Chinese agriculture relied heavily on manure, not just from livestock, but also from people (more on this topic later in the book). F. H. King, in his classic treatise "Farmers of Forty Centuries," describes in detail the way early farmers carried out these practices. Cornell University has collected numerous other examples of how ancient peoples used manure as a soil amendment (see Resources).

Wherever people have kept livestock, they have had to deal with the stuff that accumulates behind or underneath them. Early farmers must have noticed that putting manure back on the land not only disposed of it, but also dramatically improved the yield and quality of their crops or pastures. What started out as an exercise in waste disposal transformed into an agronomic practice.

In our modern age, with the advent of chemical-intensive, industrial-scale farming, some farmers seem to have reverted to disposing of manure as a waste product. In some cases, farms even dispose of manure by burning it.

If you're reading this book, I doubt you look at manure as a liability. For the past decade or two, the demographics of agriculture have been changing in the United States, and the number of small farms has increased dramatically. Many of these smaller operations embrace the concept of sustainability, and most organic farms are small farms, by today's measures of scale. We recognize the value of animal manure, not only as an economical soil amendment, but as an essential part of the cycle of life.

CHAPTER ONE

CHARACTERISTICS OF MANURE

The Nature and Properties of Crap*

The physical, chemical, and biological properties of manure make it a truly amazing soil amendment. Depending on the species of animal, the type of bedding used, and how manure is stored and spread, some type of manure is an appropriate soil amendment for almost every crop grown in almost any type of soil, anywhere in the world.

* With apologies to Dr. Nyle C. Brady, original author of the excellent introductory textbook on soils with a very similar title.

To make the best use of manure, it helps to understand its properties, along with the requirements of the crops you're growing and the characteristics of your own soils.

PHYSICAL PROPERTIES

Fresh manure is usually moist, its consistency ranging from firm and well formed to almost a slurry. Depending on the animal it comes from and what that animal eats, it may have a mild, earthy odor or a strong and unappealing smell as it is excreted.

How manure is collected and stored and whether or not bedding is included affect its physical characteristics as well as its nutrient value. Manure that has been composted doesn't even resemble the original material, being well mixed, mellow, and pleasant to the smell. Manure that is stored as a liquid in a lagoon or holding tank doesn't resemble the original material, either, but usually has a powerful odor and requires special equipment to deal with it.

Because small-scale farmers and gardeners most often will be working with solid manure, that's what most of this book will focus on.

Manure Production

It's handy to know how much manure you can expect to get from whatever kind of livestock you keep. If you consider the weight of the animal, there's a fairly narrow range of how much dry matter (the weight of the material without any water) different livestock species excrete in their manure. For each pound of live animal weight, cattle produce the least — around 3¾ pounds (1.5 kg) per year; horses, sheep, and swine around 4 pounds per year; and poultry around 4¼ pounds (1.9 kg) per year.

If you consider the actual fresh weight of the manure (including moisture), however, the range is much wider: poultry, sheep, and horses produce between 11 and 12 pounds (5–5.4 kg) of actual fresh manure per pound of live weight each year, while cattle produce 26 to 38 pounds (11.8–17.2 kg) and swine around 25 to 26 pounds (11.3–11.8 kg) per year for each pound of live weight. There's quite a range in the moisture content of manure excreted by the various species (see chart). In addition, some types of livestock manure almost always come with some bedding for good measure.

The physical (and chemical) characteristics of manure vary by species, bedding types, and storage and handling systems. The species of livestock the manure comes from is the starting point and has the most influence, so it's helpful to understand the differences.

Manure Production Stats

Here are some of the ranges in moisture and overall annual production for fresh manure (not including bedding) from several types of livestock. *Note:* the weight of bedding can be substantial.

Per Year

ANIMAL	MOISTURE CONTENT OF MANURE (%)	YEARLY MOIST MANURE PRODUCTION
Horses (1,000–lb/450–kg animal)	66%	9 tons
Dairy cattle (1,400–lb/635–kg cow)	80–85%	27 tons (25 t)
Beef cattle (1,100–lb/500–kg cow)	80%	15 tons (13.6 t)
Sheep, goats (150–lb/70–kg avg.)	65–75%	0.5–1 ton (5–10 t)
Swine (150–lb/70–kg feeder pig)	85%	1.7 tons (1.5 t)
Poultry (5–lb/2.4–kg bird)	60–65%	160 lb (73 kg)
Rabbits (8–10 lb/3.6-4.5–kg)	16%	38–48 lb (17–22 kg)

Comparison of Fresh and Dry Manure per Pound* of Animal Waste

ANIMAL	FRESH MANURE (LB)	DRY MATTER (LB)
Horses	15–20	4
Cattle	25–40	3¾
Sheep	18	4
Swine	22–26	4
Poultry	11–25	4¼

* See page 105 for Metric Conversion Table.

Cattle dung can be either cakey (when cattle are fed low-quality hay or graze overmature pastures) or very runny (in high-producing dairy cattle or any livestock grazing on lush pastures or other extremely high-quality forages).

Small ruminants (sheep, goats, and rabbits) usually have nicely formed, relatively dry, pelletized feces.

Horses tend to make bigger "balls" of manure (sometimes called horse apples).

Swine have stools that are poorly formed and quite wet.

Poultry manure is a bit drier, but it's very concentrated and is a mixture of both solid and liquid waste, since birds don't urinate.

CHEMICAL CHARACTERISTICS

ASIDE FROM THE NUTRIENTS that leave the animal (or the farm) in the form of meat, milk, or fiber, almost everything livestock eat gets passed along in their feces and urine. On average, an animal excretes around 70 to 80 percent of the nitrogen, 60 to 85 percent of the phosphorus, and 80 to 90 percent of the potassium in the feed it eats. Since most of these animals' diet comes from plants, this is a great opportunity to recycle the nutrients needed for producing more crops. And this recycling isn't limited only to the major plant nutrients of nitrogen, phosphorus, and potassium; it also includes minor and micronutrients essential for plant growth.

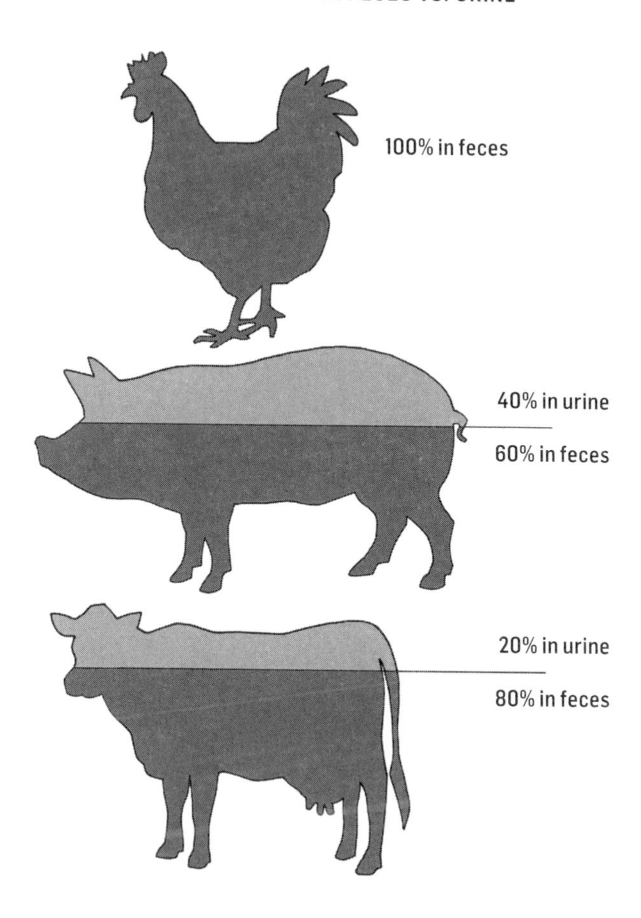

DIFFERENT ANIMAL SPECIES have very different proportions of nitrogen in their feces and urine.

Number One or Number Two?

The solid part of manure (dung or feces) differs greatly from the liquid portion (urine) in its chemical properties. In order for plants to use them, nutrients must be in a soluble form. Manure holds both soluble (available) and decomposable (slow-release) plant nutrients.

Not all the feed an animal eats is completely digested as it passes through the animal. Because the undigested solids that pass through an animal are contained in the feces, it stands to reason that it will take longer for any nutrients in them to decompose in the soil before the plant nutrients they contain can become useful to crops. Urine, on the other hand, only contains substances that are completely dissolved, so most of the nutrients in this liquid are available almost immediately for plants to use. This combination of quickly available and slowly available plant nutrients gives manure the ability to provide both short-term and slow-release plant nutrients.

The proportions of urine and feces in excrement, and their chemical characteristics, also vary by species. Poultry excrete everything as feces. Cattle and horses excrete around 80 percent of their output as feces; for sheep it's about 67 percent; and for swine, about 60 percent.

Chemically, the proportions of the "big three" plant nutrients also vary by species. Here's what it looks like on average for four species.

Approximate Percent of Nutrients in Feces/Urine

SPECIES	NITROGEN	PHOSPHORUS	POTASSIUM
Cattle	49/51	100/0	15/85
Sheep	63/37	95/5	30/70
Horse	62/38	100/0	56/44
Swine	67/33	43/57	57/43

Adapted from Van Slyke (1949) and Brady and Weil (2007)

The proportions of solids and liquids and their respective nutrient contents are important, because the system you use to collect the manure can influence the nutrients in the stored product. If the collection and storage system you have catches and stores the urine, you'll have a better product than if the urine is lost.

For example, if a farmer scrapes up cattle dung from a holding area or feeding pad, the resulting manure will have much less potassium than in a contained system (stall barn, freestall barn, bedded pack, compost barn, etc.) that catches both the urine and the dung, because most of the potassium that cows excrete is in the urine.

MANURE AS A FERTILIZER

MANURE IS AN EXCELLENT source of plant nutrients, because most of what animals eat comes from plants, and most of what the animals eat gets passed through in the excrement. All plants require a certain number of elements for their growth, and some plants benefit from a few others.

It's Elemental

Besides the nutrients that plants get from air and water (carbon, hydrogen, and oxygen), there are at least 14 **essential elements** that all plants absolutely need in order to live: nitrogen, phosphorus, potassium, calcium, magnesium, sulfur, boron, chlorine, iron, manganese, zinc, copper, molybdenum, and nickel. In addition to these necessary elements, a handful of others may benefit some or all plants, but they aren't universally required: silicon, sodium, cobalt, and selenium. Most plant scientists now refer to this group as **beneficial elements**.

The number of elements considered either essential or beneficial has grown in my lifetime and probably will continue to. Other elements have been proposed for one or the other of these lists, including chromium, vanadium, and titanium.

Other Citizens of the Soil

Healthy crops arise from healthy soils, and the soil community depends on much more than just plants. There are many other types of organisms living in the soil that work together to make a healthy, functional soil environment. Bacteria, **actinomycetes**, fungi, protozoans, nematodes, arthropods, earthworms, and other groups of living things all inhabit the soil and perform jobs that help the whole system work well. These organisms, as well as livestock and people, need nutrients beyond what just the plants require. Some other elements that may be required by microbes, animals, and people include bromine, lithium, strontium, tungsten, cadmium, and even arsenic.

Plants are the foundation of our food, even if we also eat meat, fish, or other animal products. When we apply manure to the land we raise our food or feed on, it supplies a portion of all the nutrients plants need to grow, along with most of what the livestock or people who eat those crops need.

Using Manure

Ideally, fertilizer made from manure can:

1. provide enough soluble, readily available nutrients to get a new crop off to a good start.
2. offer enough slowly available nutrients to keep the crop growing all through the season.
3. not lose many nutrients through leaching or other means.

How well that theory plays out in practice depends on the characteristics of the manure, how it's handled and applied, the weather, soil characteristics, and the needs of the crop.

When farmers apply manure consistently over a long time, it can make very long-lasting changes in the soil's fertility. At the Rothamsted Experimental Station in England, manure was applied to a plot of land every year for 19 years beginning in 1852, and then no more manure was applied. More than 100 years later, soils in the plots that received the manure still provided more nitrogen than soils in the other plots.

While it's usually possible to use manure to supply all the nutrients needed for good crop growth, that's not always the best choice. This is because manure contains nutrients that are usually available in proportions that are different from what the crops need to grow well. If we apply manure at rates high enough to satisfy all the nutrient requirements of the most limiting element, we usually apply a lot more of other nutrients.

In some cases, that doesn't cause any problems, but in others it can. In chapter 3, we'll see some examples of how this works.

BIOLOGICAL CHARACTERISTICS

Besides its nutrient content, manure by its very nature is a source of fresh **organic matter**, which is almost always beneficial for soils. The term refers to any material that is or was part of a living organism, including plant stems, leaves, roots, bark, soil animals like insects and earthworms, and microscopic creatures like bacteria, actinomycetes, fungi, and nematodes. It also includes the chemical substances that these organisms produce, such as proteins, sugars, fiber, and oils. All of these things — alive, dead and decomposing, or completely decomposed — constitute organic matter.

Organic Matter and the Soil

Soils are vibrant and complex communities of many different organisms, all of whom have a job to do. Soil ecologists estimate that there can be more living organisms in one gram (about a teaspoon) of healthy soil than there are people on Earth — and that's more than seven billion! Fresh organic matter (such as manure) serves as the food source for all these living things except plants (which produce organic matter from the elements in the air, water, and soil). As soil organisms consume fresh or decomposing organic matter, they release many of the nutrients in the organic matter and turn it into forms that can be taken up again by plants.

Soil Structure

As these organisms decompose organic matter (or live in a mutually beneficial association with living plants), they also enhance the structure (**aggregation**) of the soil. When a soil

Good Guys and Bad Guys

Not all of the organisms living in the soil are helpful for the crops we want to grow. Also in the mix are organisms that cause fungal, bacterial, and viral diseases; nematodes and insects that feed on crop plants; and other bad guys. In a healthy and well-balanced system, however, the good guys generally win.

is aggregated, the individual particles of sand, silt, and clay group together into structures called **aggregates**, or "**peds**." Earthworms are famous for their activity in forming aggregates. Other organisms, particularly the types of beneficial mycorrhizal fungi that colonize the roots of most plants, produce compounds that keep aggregates glued together.

Each of these aggregates acts like a little piggy bank to hold moisture, humus, and nutrients that help plants to grow well. Between these aggregates are small channels or pores that allow air and water to infiltrate the soil and move through it. These pores also allow microbes to move through the soil and help plant roots penetrate the soil easily so they can develop fully. All these things work together to help bring water and nutrients to the plants so they can thrive.

Humus

The residues produced by these organisms and the remaining organic matter that can't be used as food by anything else that lives in the soil become the substance we refer to as **humus**. Humus can hold onto and exchange nutrients and water that plants need. It also helps hold soil aggregates together, and it's the substance that gives topsoils their dark color. Humus can take many years to form, and only a tiny fraction of the original organic matter that enters the soil becomes humus. The rest is digested (oxidized) into water and carbon dioxide or becomes nutrients that are used by plants and other organisms in the soil. Thus it takes a huge amount of fresh organic matter to make the tiny amount of true humus in most soils.

All forms of organic matter, from the recently dead organism to the most stable humus, are very important to healthy and productive soils. Manure is one excellent source of this valuable organic matter.

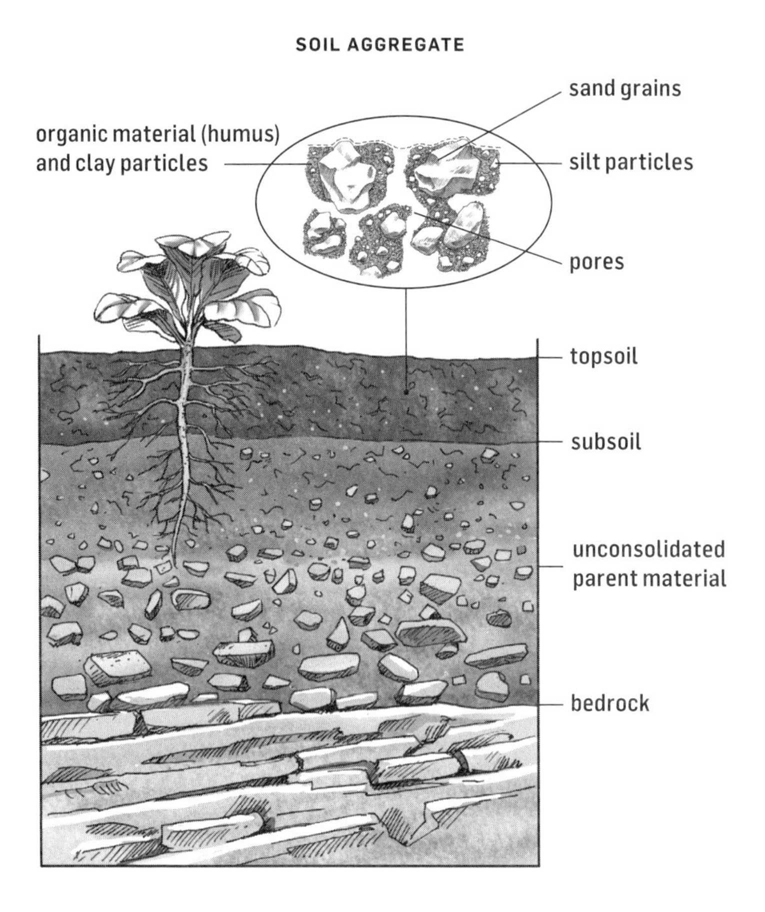

PROBLEMS WITH MANURE

While manure is generally great for amending soil, it can also introduce some problems.

Too Much Carbon or Not Enough

When too much carbon-rich bedding like sawdust, wood shavings, or straw is mixed in with the manure, it can cause an imbalance in the carbon-to-nitrogen (C:N) ratio (see below). Most plants, except for legumes whose **rhyzobial** bacterial friends take care of their nitrogen needs (see box on page 22), will grow very poorly in these situations. If you find that plants

---

Carbon-to-Nitrogen Ratio

The soil organisms that break down organic matter want a diet with a balance between carbon (C) and nitrogen (N). Soil bacteria use the carbon for their energy source and the nitrogen for the protein in their tissues. When the C:N ratio is in the range of 20–30:1, decomposition proceeds quickly and efficiently. If the ratio is higher on the carbon side, the soil bacteria will steal nitrogen that plants would otherwise use for their growth. Bacteria always outcompete plants for nitrogen. They won't release it again until they're finished with their meal — days, weeks, or even months later, depending on the characteristics of the residue they are eating. At that point, the nitrogen they have scavenged can be released again and plants can grow normally.

---

look pale or yellow after you apply manure to their bed, most likely there is too much carbon-rich bedding for the amount of nitrogen in the manure.

Many types of manure have a lot of nitrogen compared to the amount of carbon if no bedding is added. You can still use this type of material, but it behaves more like a fertilizer than manure with bedding, which we usually think of as a soil amendment. If you apply manure this way, it's important to consider the nutrient content, especially for nitrogen, and not apply it at rates that are higher than the demand for the crop. Applying more nitrogen than what the crop needs can lead to weak plant stems, failure to fruit, or poor-quality vegetables.

Remedies. If there is more nitrogen in the manure than what you want to apply at the rate you want to spread, add more bedding to bring the nutrient concentration down. If there isn't enough nitrogen to balance out the carbon in the bedding, add more bedding-free manure to bring the carbon-to-nitrogen ration into a better balance.

You can also compost the manure or let it decompose naturally in a pile before you apply it. This usually works better with high-carbon manure and bedding mixes than with high-nitrogen materials. With an excess of carbon, the main problem for **composting** is that it may take a long time. When there is too much nitrogen, the pile can easily become anaerobic, smelly, and runny.

When most of the decomposition is done, the bacterial process is nearly complete and the manure can be applied without any ill effect from inadequate nitrogen.

Nitrogen Fixation

Nitrogen is almost always a limiting nutrient for maximum plant growth. Nitrogen is found in hardly any rocks and minerals, so it doesn't weather out of the parent materials soils are made of. The atmosphere, on the other hand, is teeming with nitrogen: around 78 percent of the air we breathe is nitrogen, but it's in a form that plants can't use. Fortunately, some types of microbes come equipped with an enzyme called **nitrogenase** that helps them convert atmospheric N into a form they can use for their own growth, a process called **biological nitrogen fixation**.

This is a very energy-intensive process, and these microbes rely on host plants to provide the energy, furnishing the plant with nitrogen in return. This symbiotic relationship is prevalent in the legume (pea and bean) family of plants, and the group of bacteria that work with these plants is called *Rhyzobium*.

Chemical Contaminants

Depending on where the manure comes from, it may also have chemical contaminants. Hormones and antibiotics used in some conventional livestock operations can be passed through the manure and end up contaminating the soil. The same thing can happen with insecticides and **anthelmintics** (deworming chemicals). Some of these substances don't cause direct harm, and some are broken down into fairly benign compounds by the soil biology, but others can negatively affect the organisms in the soil. For instance, a popular type of dewormer, ivermectin, can pass through the livestock and inhibit the dung beetles and other insects that are the first in line to digest the manure when it's applied to the soil.

If livestock bedding is contaminated with herbicides, some of these residues can also be present in the mix that gets applied to the soil. It's not common, but a few of the herbicides used today, like aminopyralid, are exceptionally resistant to the decomposition process and can cause problems for the crops where the manure is applied.

The sanitizing chemicals used to clean milking equipment can be a problem if milk-house waste is mixed with the manure, because these chemicals are designed to kill bacteria and fungi like the ones we want to feed in the soil. Even so, these chemicals typically are diluted by huge volumes of manure and become essentially a drop in the bucket.

Odor

Odor isn't usually a problem for crops, but it can be a problem for your neighbors. Usually pit manure causes more of an issue than piled or fresh stall manure.

Remedies. Apply manure when the wind is blowing away from your neighbors, and quickly incorporate the manure into the soil after you spread it.

With solid manure, maintaining a well-aerated composting process or leaving an undisturbed **static pile** with a good balance of carbon-rich bedding in the manure usually takes care of most of these problems. A strong odor of ammonia can be a sign that there isn't enough carbon-rich material, and adding some bedding (even over the top of the pile) will usually resolve the situation.

CATS, DOGS, AND PEOPLE

WHAT ABOUT ALL THE WASTE from our pets, and from ourselves, for that matter? There's no doubt that every day we and our pets generate lots of excrement, which is loaded with organic matter and nutrients. In Asia, agriculture functioned for thousands of years using "night soil" (untreated human excrement) as a vital source of nutrients to grow all sorts of crops, including vegetables.

In many areas growers have discontinued this practice on food crops, and that's a good thing. Human waste contains pathogenic viruses, microbes, and parasites, so it should not be used on crops grown for food. Dogs and cats also can harbor many nasty microorganisms that can infect people, so it's not a good idea to use excrement from any of these species on any crops for our food, and probably not for livestock feed, either. (Even with livestock manure, you shouldn't apply it to vegetable crops less than 90 days before you harvest.)

Remedies. We don't need to waste these resources, however. Pet manure can fertilize ornamental trees and shrubs in areas where people don't normally go and kids aren't likely to play. Cover the material with some kind of high-carbon mulch (wood chips, bark, etc.) for good sanitation, to keep odors down, and to discourage animals from visiting the area. This also provides a good carbon source to help balance out the nutrient supply to the microbes, earthworms, and other organisms that will break down the material.

Humanure

Human waste is a bit trickier. Few people would want their own fresh leftovers placed around the base of the shrubbery in their yards, even with a good mulch cover. Many municipal sewage facilities, however, treat sludge from their sanitary plants and process it for use as a fertilizer. When sewage is treated this way, the pathogens are disabled and the material can be used safely, from the standpoint of infectious disease and parasite contamination.

Despite that, human waste, or **humanure**, is best used on nonfood and nonfeed crops. Here are some reasons.

Industrial and household chemicals contaminating municipal sewage supplies pose a problem. Treated sludge is tested for its heavy metal content, which must be below the legal limits before it can be sold as a fertilizer, but many people would prefer not to put any of this material on their food crops. Unless the material is coming from your own home, there's really no way of knowing what kind of household chemicals people are pouring down their drains.

Pharmaceutical contamination of human waste may be an even bigger concern. A growing body of scientific literature is documenting the presence of antibiotics, painkillers, and hormones in human waste. Many can be degraded through appropriate treatment, but some persist in the environment.

Hormones influence biological systems even at incredibly low concentrations — at the parts per billion level. That's a hard concept to visualize, so here is an analogy. One part per billion is equivalent to 1 inch in 16,000 miles (two-thirds of the way around the earth), or 1 second in 31½ years.

The prevalence of these substances in wastes, and even in ground- and surface water, is alarming. Considering that hormones cause reactions in organisms at such minuscule levels, it seems like a good idea to avoid using any type of waste that we aren't sure of with respect to these contaminants.

ONE POSSIBLE EXCEPTION: HUMAN URINE

Human urine is an excellent source of plant nutrients, including nitrogen, phosphorus, potassium, sulfur, and small amounts of micronutrients, and is close to sterile when it comes from healthy people. Despite a fair amount of sodium, it can be an excellent fertilizer for many plants. When its source is healthy people who are not taking hormone treatments or other persistent medications, it can be used safely even on food crops. Toilets can be designed to separate the solids from the liquids, and even in your basic outhouse, it's easy to segregate that part of at least the men's activity from the rest of the material.

GETTING IT RIGHT

To learn more about using human excrement as a fertilizer, read *The Humanure Handbook* by Joseph Jenkins (see Resources).

CHAPTER TWO

STORING AND HANDLING MANURE

If you have livestock, you need a system for storing and handling the manure your animals generate. The common types of manure-handling systems are solid (either daily haul or stored), semisolid or **slurry**, and liquid. The type of manure-storage system you have may be determined by the type of facility that was on your farm when you bought it, or you may decide that a different type of system is more appropriate for the size of your farm or your management practices.

Solid manure systems may be as simple as a pile of manure on a slab or in an appropriate location on the ground, or as complex as an engineered structure specifically designed to protect the manure from the weather and to make it easy to work with. In one type of solid manure system, the **bedded pack**, the animals are fed and bedded in the same area for a fairly long time so that a mound of bedding, manure, and waste feed accumulates. When you move the animals out of this combined housing and feeding area, either collect and spread the manure as is, or scrape it up to compost or simply rot over time.

Semisolid systems are usually in facilities that have a shallow concrete-lined basin with a treated wood "picket fence" end wall. These systems catch both manure and wastewater, and the end wall is designed to let excess liquids flow out to a vegetated buffer strip or to another containment facility just for liquids that can used for irrigation. The material in the main collection area is a soupy mix of manure and liquids, spread most easily with a slinger-type spreader that doesn't leak liquids on the way to the field.

Liquid manure systems collect manure and wastewater the same way the semisolid systems do, but they retain all of this material in engineered, lined pits or aboveground reservoirs. You need to physically mix (agitate) them before pumping them into liquid manure spreaders for application. Liquid manure spreaders can either spray the material over the ground or inject it directly into the ground.

Because of the specialized storage facilities and related equipment that semisolid and liquid systems require, they are less common on small livestock farms. This book will focus on solid manure storage systems, which offer several options for handling the manure.

SOLID MANURE HANDLING SYSTEMS

You can collect and spread solid manure every day (the "daily haul"), or collect and store it for later spreading, using a range of implements. Each has advantages and constraints.

Daily Haul

TYPE OF LIVESTOCK: Typically dairy cattle or other highly managed livestock that generate a lot of manure in a fairly small area.

HOW IT WORKS: Daily spreading of manure (and associated bedding) produced by a herd that is housed inside.

PROS: No storage needed (except briefly during inclement weather). Minimal nutrient losses from the original material. Odor from spreading may be intense in the immediate area but usually dissipates quickly. Nutrients are available to the crops quickly after the manure is produced.

CONS: Labor intensive; manure will need storage when the ground is too soft for transport; requires a cropping system that can have manure applied year-round. Almost all weed seeds in the original manure remain viable. Livestock of the same species will refuse to graze pastures applied with fresh manure, usually for several months. Excessive application rates can cause physical damage to crops by salt injury. Daily spreading can be hard to do when the snow gets deep or when the ground is soft from excessive rain or while the frost is coming out of the ground in the spring.

HOW TO IMPLEMENT: Collect the manure either by hand or with a barn cleaner, load it in the spreader, and apply it directly to a field.

EQUIPMENT NEEDED (not all apply for every situation): Scraper or pitchfork, wheelbarrow, barn cleaner, tractor and loader, manure spreader.

GETTING IT DONE

Many family-sized dairy farms employ the **daily haul system**. Each day they spread any manure (and associated bedding) produced by a herd while it's housed inside.

This system has the advantage of not requiring any substantial storage capacity for the manure (except for brief periods of inclement weather), which can save some money on facilities, but it also has some down sides. It means one more set of daily chores. If transportation and spreading are a problem, as when the ground is soft from a lot of rain or deep in snow, some short-term storage is required.

The farmer must be able to get the manure to the fields that will benefit the most from it. The distance you have to haul is a factor for most manure systems, but especially important with daily haul systems. In fact, there is a common pattern of nutrient distribution on most farms with a long dairy history. The fields that are close to the barn typically have very high levels of fertility (especially phosphorus, and to a lesser extent potassium). As you get farther from the barn, the fertility almost always drops off. This is due to generations of daily haul manure spreading, where the farmers took the manure only as far as necessary to spread it. These close fields usually need little extra fertility, while the more distant fields are often starved for nutrients.

Solid Storage (*Static Pile*)

TYPE OF LIVESTOCK: Any

HOW IT WORKS: Manure is collected and piled without any further processing until the time comes to spread it.

PROS: Simple, requires minimal equipment. If the manure is left in a pile for a long time (over a year) it will slowly decompose, reducing the volume and making it easier to handle. Some weed seeds die during the storage time. If the manure is well rotted, it can be applied to pastures without causing grazing livestock to refuse the forage.

CONS: Requires a storage area that will protect the pile from runoff or leaching losses. Many weed seeds remain viable during storage. There will be nitrogen losses during storage, and potassium will leach out if the pile isn't protected from leaching or runoff losses. It requires a long time between when the manure is produced and when it can be spread.

HOW TO IMPLEMENT: Find an appropriate place to pile the manure, and start collecting. As you collect manure over time, extend the pile in a row or start a new pile so that you can eventually spread the rotted manure without disturbing the freshest material.

EQUIPMENT NEEDED: Same as for daily haul systems.

GETTING IT DONE

Piling manure isn't rocket science, but there are still a few things to keep in mind to conserve the nutrients in the manure and to be a good environmental steward.

Location, location, location. For starters, manure should never be piled where it can either leach into the soil or wash away and contaminate surface waters (lakes, streams, ponds, etc.) or groundwater (which is drinking water for most of us). This means choosing a spot that's fairly level and not in the path of surface runoff from heavy rains or melting snows. Runoff that erodes the pile or moves some of the manure is obviously a problem, and most people know better than to put a manure pile in harm's way.

Nutrient loss. Less obvious is the substantial nutrient loss that occurs when rainfall or snowmelt leaches *through* a pile of manure. In particular, nitrogen and potassium are both quite soluble in manure and can easily be lost through leaching. Both are often in short supply in the soil, and expensive to buy as fertilizer, so it pays to conserve them. Some nitrogen is always lost in the form of ammonia through volatilization, but there are several things you can do to minimize these losses. Having enough carbonaceous bedding mixed with the manure helps. For static piles, it's actually better to have lots of moisture in the manure and to keep from disturbing the pile, to keep it in an anaerobic state. With this method, the manure takes a long time to break down, but the nutrients are conserved quite well.

Protecting groundwater. With regard to groundwater, nitrogen can be a problem because both ammonium and nitrate are

Pastures Need a Waiting Period

If the area that needs the nutrients from manure is a pasture, there's another problem with daily haul systems. Fresh manure spread on a pasture means that it's going to be a while before livestock will graze on that part of the land. Whether it's the smell of the fresh manure itself, or some built-in defense mechanism in the brains of the livestock that helps them defend against parasites, no animals will willingly graze on pastures that have had fresh manure from their own species applied recently. The amount of time they will avoid these areas varies depending on the application rate, the amount of rain, and the temperature. It may be only a couple of months, or it could be a lot longer. (See page 72 for more information.)

extremely soluble. The nitrate form of nitrogen is not held very well in the soil and can easily percolate through to the water table. Excessive nitrate levels can cause big problems for infants and pregnant women, and they don't do the rest of us much good either. In some areas with shallow water tables or fractured limestone bedrock, even the bacteria from a manure pile can get into groundwater and contaminate drinking water.

Between the potential for crop nutrient losses and the risk to the environment, it makes sense to prevent too much water from getting into manure piles. In some climates, it is best to keep manure piles under a roof, or at least a tarp, to keep this from happening.

If you use an uncovered bedded pack system for your livestock, push the pack into a pile when you stop using it. This will put the total volume of manure into a smaller footprint, so that precipitation that lands on the pile will have a greater depth of material to work through before it can leach.

When you make a pile, be sure to mix the manure and bedding as well as possible. The mixture should be moist, but not runny. Make the pile as compact as possible, and try to keep the sides fairly smooth to reduce the amount of surface area. This will help conserve the nutrients in the manure and will cut down on odors until you next work with the pile.

Solid Storage (*Compost*)

TYPE OF LIVESTOCK: All

HOW IT WORKS: Manure is piled up, usually along with a high-carbon type of bedding, and mixed frequently (aerated) to speed up the decomposition process and heat the pile.

PROS: The end product is loose and friable and contains very few viable weed seeds. Most pathogens (both plant and animal) are killed. Few weed seeds survive a properly managed, hot composting process. Nutrient loss is minimized through good **composting** practices. Nutrients are in a stable form that makes an excellent slow-release type of soil amendment. The only odor is a mild, earthy scent. Livestock will not refuse to graze on pastures that have had compost applied recently. Nutrients are available to the crops within a few weeks or months from the time the manure is produced. There is very little potential to injure plants by applying compost unless they are physically smothered by a mulching effect. The volume of the material is greatly reduced during composting, so not only is it easy to spread, there's only about one third as much material as you started with.

CONS: Time-, management-, and likely equipment-intensive. Small quantities of manure can be composted by mixing by hand; otherwise it will require mechanical equipment. Compost is applied at much lower rates than ordinary manure, so spreading it may require a different type of spreader from the common ones used on farms. Composting is difficult to do in the winter in areas where the weather gets very cold.

FOR SMALL AMOUNTS of manure, a compost bin system works well.

HOW TO IMPLEMENT: Composting is done most effectively on a concrete pad if you will use a tractor or skid steer and loader. Small batches that are turned by hand or big operations that use a mechanical compost turner can use well-drained, firm, and level ground.

EQUIPMENT NEEDED (depending on the scale): Pitchfork, wheelbarrow, water hose (to add water in dry areas), tarps (to keep excessive rainfall out), tractor or skid steer and loader, compost windrow turner, compost spreader.

GETTING IT DONE

Composting manure means a lot more management and possibly extra equipment, but it offers some real advantages over simply piling manure.

There are good books available that describe composting in great detail, but the basics are fairly simple. For a good composting system, the C:N ratio of the mix of manure and bedding is very important. This usually means a mixture of bedding and manure. The mix needs to be sort of fluffy, indicating that there is room for air in the pile, so the pile needs to include some fibers that are long enough to serve as bulking agents without being so big that they won't compost. There needs to be enough water in it to be moist, but not wet.

Once you have the right mix of materials at the right moisture, make a pile or a windrow that you can turn frequently to introduce fresh air and release carbon dioxide. The pile should heat up rapidly (within the first day of assembling it) and should get so hot on the inside that you won't be able to hold your hand in it.

Actual compost for organic farming systems has to be made in accordance with the rules established under the **National Organic Program (NOP)**. If it doesn't meet all these requirements, it may behave similar to "official" compost, but it has to be treated as ordinary manure when it comes to organic farming rules. For more information on these requirements, see pages 87–88.

LARGER AMOUNTS OF MANURE can be piled in rows to make it easier to turn for composting.

In a typical mechanical composting system, a pile of manure and bedding in the correct proportions is turned or mixed thoroughly and frequently. You can turn and mix small piles by hand with a fork, but most people use a tractor or skid loader. Larger composting farms use a mechanical windrow turner that travels down the length of a windrow and turns and mixes the materials as it proceeds. This type of equipment is probably not practical for most small farms unless they have a market for finished compost and can get other organic materials to compost, because the investment in the machine calls for producing a lot of compost to make the process pay.

Whatever type of system you use, the principles are the same.

Maintain a healthy brown-green balance. You need the right combination (see below) of **carbonaceous** ("brown") material and **nitrogenous** ("green") material. Most manure has more than enough nitrogen and needs some extra carbon (usually bedding of some sort) for good composting, and that's why it's considered "green" — though it often looks brown.

The ratio of brown to green material will vary depending on the characteristics of each of the materials, but the C:N ratio of the starting material should be in the range of 30:1. For small compost piles that will be used on the farm, it isn't as critical for this ratio to be in this range as it is with large operations or when you're trying to make USDA Organic certified compost. In those cases, you should consult the National Organic Program's composting rules to understand all the requirements, including time and temperature specifications.

TURNING A MANURE PILE BY HAND. To protect the muscles in the back, use a long-handled tool, ideally with an ergonomic handle (shown), and stand as upright as possible. Change your position and grip frequently to prevent repetitive stress.

Provide moisture and aeration. The material must be kept moist (but not too wet — see page 33), and it requires a constant source of oxygen for microbes to digest it quickly.

Turn and mix. The organisms that do the work in a compost pile are mainly **aerobic** — in other words, they breathe in oxygen and breathe out carbon dioxide and water vapor, just as we do. Turning or mixing the compost pile allows carbon dioxide and water vapor to be released from the pile and brings fresh oxygen in.

A TRACTOR AND LOADER come in handy when working with a large quantity of manure.

HEATING UP

During this active composting process, the microbes that are decomposing the manure and bedding generate a lot of heat. An intensively managed composting system maintains this heat at a high enough temperature, for a long enough time, to kill weed seeds and pathogenic organisms. If you're making compost according to the USDA standards, you must actually measure the temperature of the material with a probe-type thermometer. If you're just doing it informally, you'll know it's working right when you can't hold your hand inside the pile because of the heat.

These microbes also produce compounds that help crop plants resist some plant diseases. Amazingly, the microbes can withstand the heat of the active composting process, which pasteurizes the product, without getting killed themselves. The material produced this way is the only material that's considered true compost, under the National Organic Program's (NOP) standards. For more details on the exact requirement, consult the USDA standards at www.ams.usda.gov/nop.

For everyday composting, if you want actual compost with all the characteristics outlined earlier, the pile must heat up substantially and decompose aerobically. If the pile isn't heating up, you may need to add some water, add more green materials, use materials with smaller particle size, or turn the pile more frequently. If the pile starts to smell foul (with a manure-like ammonia smell) or seeps water, add more brown material, preferably with less moisture.

Manure that is just left in a pile and not mixed will also eventually decompose. This type of well-rotted manure has some of the same physical and plant nutrient characteristics that true compost does, but it can still harbor weed seeds and may contain pathogenic organisms.

DON'T BE ALARMED when a manure pile starts to let off steam. It's a sign that the fermentation (composting) process is in full gear.

Compost Barn

TYPE OF LIVESTOCK: Dairy animals

HOW IT WORKS: In this system, you spread a thick layer of high-carbon bedding, such as sawdust or fine wood shavings, over the whole floor of the housing area. Each day, you mix the manure deposited by the livestock into this bedding, usually with a skid loader or small tractor with a cultivator-type attachment. From time to time you add more bedding to keep the animals clean.

This combination of high-carbon bedding, high-nitrogen manure, moisture, and air causes the bedding pack to actively compost while the herd uses it. This system also generates heat, which in areas with cold winters can make a substantial difference in the temperature of the loose housing area. When you finally clean out the pack, much of the material has already been composted and that portion has the same characteristics and advantages as true compost made by piling and mixing.

PROS: Animals stay very clean, and udder health is excellent. The material is mostly composted and has many of the same characteristics that ordinary compost does. This is a system that actually works well for small farms, if you can afford the investment for the facilities and bedding and have a consistent supply of sawdust available.

CONS: This system requires large amounts of high-quality bedding that is available consistently — this is a constraint in many areas. The housing area needs to be indoors in an area that will accommodate the machine used to stir the pack, and the pack has to be stirred every day in most systems. Because loose housing demands a fairly large area for each animal, this kind of housing requires a larger footprint than other types of housing like freestalls or tie stalls.

HOW TO IMPLEMENT: Consult your university Extension agriculture professionals for design standards. Several universities have web-based publications that cover **compost barns**, and reviewing those would be a good place to start (see Resources).

EQUIPMENT NEEDED: Facility, small manure spreader for bedding, tractor or skid steer with cultivator and loader, manure spreader.

GETTING IT DONE

A layer of finely divided high-carbon bedding (usually sawdust) is applied to the floor of an indoor loose housing area. As the animals occupy the housing, a tractor or skid steer with a small field cultivator/quack-digger attachment is used (usually daily) to mix any fresh manure into the bedding. When it gets too hard to mix enough bedding into the manure to keep the animals clean, another layer of bedding is spread through the housing area and the process is repeated. When the mix of bedding and manure gets deep enough, the material begins to heat and actually compost, killing pathogens and weed seeds. When the compost pack gets so deep it starts to interfere with the operation, or when manure is needed for a crop, the material is scooped out with a loader and either stockpiled for spreading at the appropriate time or spread at that time.

Composting Livestock Carcasses

Dry manure that has an excess of carbon from a high volume of bedding can be used as the matrix or bulking agent for composting the carcasses of dead animals, from mice to workhorses. This is not active composting, where the pile is turned and aerated; it's more of a specialized application for a static pile.

Place the carcass on a deep layer of high-carbon material (straw, sawdust, wood chips, etc.) or well-rotted manure, then cover it with a very generous surrounding of more carbon-rich bedding material, or more carbon-rich manure. The carcass basically rots away inside the pile, and the liquids that seep down are held in the base material. There the carbon in the base can mix with the nitrogen from the carcass to result in a good ratio of carbon to nitrogen encouraging decomposition to continue. The material surrounding the carcass is deep enough to keep odors down by absorbing the ammonia and other gases that are evolved during the process.

This process should be left alone until you're sure it's done. In the summer months, that will take a few months, but in the winter it probably won't start until the weather warms up again, and then give it several months to proceed. The resulting product is safe to use as a soil amendment, but for the purposes of vegetable production, treat it like raw manure because it's not true compost.

This is a brief overview. See Resources for more information.

ENLISTING ANIMAL HELP

There are two other ways to compost manure without straining your back or using a lot of fancy equipment. Both of these rely on other organisms to do the work for us.

The "pigerator" system popularized by Virginia farmer Joel Salatin requires the help of some hogs. The idea of using hogs to root up a manure pack isn't new. Lucius Van Slyke described something along these lines in his 1949 book, *Fertilizers and Crop Production*. The system calls for spreading some corn on the bedding pack before each successive layer of bedding is applied. At the end of the cattle housing season, the cattle are sent back out to pasture and hogs are brought in to the pack area.

Hogs love to root around anyway, and when they figure out that there is buried treasure in the pack, they relentlessly tear it up, continually turning the residue and introducing air, allowing the manure and bedding to compost during the process. The final product from this type of system probably won't qualify as true compost under the NOP rule, but it's a pretty good material. I'd much rather load and spread this kind of manure than what usually comes out of a bedded pack.

Vermicomposting relies on worms to digest the manure and bedding. These systems require a fairly warm environment and a higher degree of management. The worm castings that result from this process are a high-value soil amendment with their own unique characteristics and are often used for horticultural applications.

PIGS AT WORK, turning a manure pack into compost.

Bedded Pack

TYPE OF LIVESTOCK: All.

HOW IT WORKS: Livestock are fed in the same area where they are bedded. As manure gets too prevalent, bedding is added. Manure, bedding, and wasted feed accumulate in one area until weather allows stock to go back out on pasture.

PROS: Simple, requires minimal equipment, and concentrates nutrients and organic matter in one area where they can be hauled from for strategic applications during good weather.

CONS: Rainfall will leach potassium, and there will inevitably be nitrogen losses. Pack manure with straw bedding is very hard to work with, whether you use hand tools or mechanical means. Depending on wind, temperature, and precipitation, animals housed on a bedded pack may be subjected to harsh conditions.

HOW TO IMPLEMENT: Select a fairly level, well-drained location that offers protection from the elements for the typical kind of weather you have. Start feeding and bedding there, and the animals will continue to use the area as long as no other feed and no better shelter are available.

EQUIPMENT NEEDED: Hay feeders and whatever equipment you need to get feed and bedding to the pack area. Depending on the scale, you'll need either hand tools or machinery to spread the pack manure or pile it up for composting or storing in a static pile.

LIQUID AND SEMISOLID MANURE STORAGE

These systems are less common on small farms, but in some parts of the country even fairly small dairy or hog farms use pits. Liquid manure and slurry offer some physical labor savings over small-scale solid manure systems, but they also limit the ways you can use the manure. Liquid manure can't be composted easily, and it usually doesn't lend itself to selling except to nearby farms. It usually has a stronger odor than solid manure when it's being spread.

If you keep liquid manure or a slurry in a confined area, be very careful when you are working in the containment facility. Liquid manure gives off hydrogen sulfide gas, which is poisonous. In addition, the air around liquid manure systems may have higher carbon dioxide and lower oxygen levels than ordinary atmospheric air. Many farmers have died in these types of facilities.

If you need to work in an enclosed liquid manure storage facility, ventilate it well before you go in, and always have someone observing you from nearby for safety. You should have a self-contained breathing apparatus like those firefighters use and a rope tied around you, if you must work in these conditions.

CONSERVING NUTRIENTS DURING STORAGE

IF YOUR MANURE SMELLS like ammonia, you're losing nitrogen. In anaerobic (liquid) manure systems, some sulfur can be lost as hydrogen sulfide. Potassium doesn't evaporate like ammonia, but it's very soluble, so it's easily lost through leaching. Except for these, the nutrients in manure are quite stable and easy to conserve during storage as long as the manure is physically protected from being washed away.

There are several ways to conserve manure's nutrient value.

- First, keep it from washing away! The very best way to store manure is on a concrete pad under a roof. Poorly placed manure piles and poorly managed pits allow runoff to carry nutrients away, which can contaminate both surface water and groundwater (often the source of drinking water for the neighborhood).
- Incorporate manure quickly after you spread it. This works well, but in reality it's practical to do this only a few times a year and with only a few crops.
- Keep enough high-carbon bedding in the manure to absorb the ammonia that's given off.
- If you store manure in piles, manage those piles to help reduce the losses. Keep out excessive moisture to prevent nitrogen and potassium from leaching away.
- Compost manure to help stabilize the nutrients it contains, and to convert some of the very soluble nutrients in fresh manure into more stable forms.
- Add amendments to help stabilize some of the nitrogen and enrich the nutrient value of the manure. See facing page.

AMENDING MANURE TO ENHANCE NUTRIENT VALUE

AMENDMENTS ARE PRODUCTS ADDED to manure to improve its nutrient characteristics. They may help protect some of the nitrogen from being lost, or they may increase the concentration of various plant nutrients.

Gypsum

The same material that constitutes wallboard, **gypsum** (hydrated calcium sulfate) is a soft mineral that is commonly used as a soil amendment in organic farming, and it's becoming more popular in conventional crop farming, as well. It is much more soluble than limestone, but less soluble than something like table salt.

When gypsum dissolves, it releases **ions** (electrically charged molecules) of calcium and sulfate. These sulfate ions team up with ammonia (or more correctly, ammonium ions) in the manure and form ammonium sulfate. Ammonium sulfate is much more stable than the other forms of ammonium (like ammonium carbonate) that are typically present in manure, which tend to escape into the air as ammonia gas. Adding gypsum to manure, therefore, helps to conserve some of the nitrogen that would otherwise be lost in the ammonia gas. At the same time, the calcium and sulfur content of the manure is enriched.

If you want to add gypsum to conserve as much nitrogen as possible, it takes quite a bit — around 100 pounds of gypsum per ton of average manure. Another way to look at this is to think of the manure as a convenient way to spread gypsum, if you were planning on doing that anyway. Gypsum as a source of sulfur for soil fertility is usually distributed at around 200 to 300 pounds (91–136 kg) of gypsum per acre, so add the appropriate amount of gypsum into the quantity of manure you spread per acre to supply this much. We usually spread cattle manure at rates much greater than this (5 to 20 tons [4.5–18 t] /acre), so if you put the amount of gypsum in the manure that's appropriate for each acre, you're using less gypsum than it would take to conserve the most nitrogen, but it will still help.

Rock Phosphate

Another product that you can mix into manure is **rock phosphate**. This is the usual phosphorus source that organic farmers use to amend soils that are low in phosphorus if they aren't able to apply enough manure to take care of it. One of the problems with using rock phosphate as a soil amendment is that it is very slowly soluble in most soils. Mixing it into manure is one way to improve its availability to the crops, and the phosphate ions that are released in this process also help stabilize the ammonium in the manure, as do the sulfate ions from gypsum.

A combination of gypsum and rock phosphate will save even more nitrogen from the manure, but you should add rock phosphate only if your soils can stand the extra phosphorus addition (remember, manure has a good amount of phosphorus in it already). This is a practice that most modern textbooks don't talk about, but some of the old agricultural references describe it nicely. (See Resources for one of my favorites, *Fertilizers and Crop Production* by Lucius L. van Slyke.)

What Not to Add

Barn lime is often part of the mix of materials that end up in stored manure. Liming is a good practice for barn hygiene, but from a chemistry standpoint it's not the best thing to add to manure. Lime is composed of calcium carbonate, or calcium carbonate plus magnesium carbonate. Either way, when lime dissolves, the carbonate ions that are released tend to drive ammonia gas out of the manure, just the opposite of the effect sulfate or phosphate has on manure. Wood ash contains similar constituents that do the same thing.

Lime and ash can both be excellent soil amendments for soils that are too acidic, but they should be applied separately and not mixed with manure to save the most nitrogen.

CHAPTER THREE

SPREADING THE WEALTH

Many people think of manure nutrient value mainly in terms of nitrogen, and it's true that manure can be a great source of economical nitrogen. From the information in the last chapter, you can see that manure also contains substantial amounts of the other macronutrients (phosphorus and potassium), the minor nutrients (calcium, magnesium, and sulfur), and the micronutrients that crops need to grow. These nutrients are all essential to getting good yields of high-quality crops and pasture, even though the amounts needed of each of them vary.

TIME TO TEST

BEFORE YOU APPLY MANURE, test your soil to see what nutrients your fields need and which fields could benefit the most from manure. Manure provides all the nutrients that plants need, but some soils need more added fertility to perform well than others do.

Soil samples can be tested through a university or state lab where you live, or there are many reputable privately owned labs. For the purpose of figuring out where to spread manure, the basic type of test that measures phosphorus, potassium, organic matter, and pH (acidity or alkalinity) is sufficient. Even so, tests that measure the minor and micronutrients are very useful. On some farms where most fields have adequate levels of phosphorus and potassium, it might be micronutrient levels that determine the best places to apply manure.

Manure is a very good source of nitrogen, but it also contains substantial quantities of phosphorus, potassium, and other nutrients. Because there are other ways of supplying nitrogen to crops (like plowed-down forage legume crops or green manure crops), I'd encourage you to consider the total nutrient needs of your various fields. You may find that the manure you have available is more valuable for fields that need more phosphorus and potassium than on fields that will grow crops that need a lot of nitrogen.

In chapter 1, I discussed the nutrient content of manure from various species of animals. Those numbers are based on average values in manure from many farms. The nutrient content of manure on your farm may be close to these values or substantially different. Average manure nutrient values are often good enough for less intensive cropping situations and for home gardens, but if you raise high-value crops or if you want to get the most value from your manure, you should also test the manure. If you have a fairly consistent feed and bedding system and your storage system doesn't change, the numbers you get from this kind of testing should be fairly similar from year to year. Many soil testing labs can also do nutrient testing on manure, and they often include both the total nutrient content and an estimate of how much of the major nutrients will be available to crops the first year after you apply it.

Whether you are testing soil or manure, it's essential to follow a good sampling procedure. There will be variations in the nutrient levels of your soil as you move across a field, as well as in the nutrient content of manure from one point in the storage system to the next. The goal in any such program is to collect representative samples that will reflect the average conditions of the whole unit. With this in mind, take lots of individual samples that you can mix together to form a good composite sample, and don't collect samples from any places that don't seem representative.

There are good publications on both soil and manure sampling, and you should familiarize yourself with these before you collect samples to send in to the lab for testing. Your university Extension service is an excellent place to look for references on these testing options and will provide suggestions that are appropriate for your region.

Interpreting Your Test Results

For both soil and manure tests, as well as fertilizer recommendations, it's important to understand how the results are expressed.

Nitrogen (N) is usually expressed in its elemental equivalent when it comes to fertilizer materials; in other words, how much pure nitrogen is in the sample if it could be separated out from the rest of the material.

Phosphorus (P) in fertilizer is usually represented by its oxide equivalent (an old convention dating back to earlier days of chemistry). The chemical formula for its oxide is P_2O_5, usually referred to as **phosphate**.

Potassium (K) in fertilizer is likewise represented by its oxide equivalent, K_2O, usually referred to as **potash**.

Soil tests may or may not show nitrogen levels, because nitrogen is difficult to measure in soils and can be present in many different forms. In fact, most of the time when we have ordinary soil samples analyzed for nitrogen, the results aren't very useful. This is because nitrogen can be present in so many forms in soils, and labs often measure only nitrate. There are valid ways to measure soil nitrate to provide useful information, but you must follow a very specific protocol to get meaningful results. Newer tests such as the Solvita biological respiration test and the Haney test give better results, and these can be performed on ordinary soil samples.

Soil tests usually give the analysis of phosphorus and potassium listed in their elemental form, because agronomists are used to interpreting soil test *results* according to numbers expressed that way. But if they include *recommendations* for applying phosphorus and potassium fertilizer, those are usually given in the oxide form.

To further complicate the matter, the elemental concentrations in soil samples can be given as either parts per million or pounds per acre! Don't worry, though — the conversion between these is pretty straightforward: an acre of completely dry average soil to a depth of about 6 inches is assumed to weigh 2 million pounds, so to convert parts per million to pounds per acre, just multiply by two. To convert pounds per acre to parts per million, divide by two. Easy!

When fertilizer is sold, the analysis is always listed with a set of three numbers according to this formula: Percent nitrogen as its elemental basis (N)-percent phosphate (P_2O_5)-percent potash (K_2O). For example, a fertilizer with an analysis of 9-23-30 would translate as 9 percent elemental nitrogen, 23 percent phosphate (P_2O_5), and 30 percent potash (K_2O).

With manure, the nutrient value is usually expressed as pounds of nutrients per ton of manure (for solid manure) or per 1,000 gallons (for liquid). Although manure contains all the nutrients necessary for plants to grow, labs don't usually test it for anything beyond nitrogen, phosphorus, and potassium (although they occasionally test for sulfur). Manure tests are designed to measure how much fertilizer value the manure has with respect to these major nutrients. Manure test results may report the content of the major nutrients in their elemental form, as their oxide form (for phosphorus and potassium), or both.

When you're calculating how much manure to apply in order to meet the recommendations on your soil test report, it's usually easiest to use the oxide basis to compare the nutrient value of the manure to the amount of fertilizer it would take to supply the same amount of nutrients. Note: Some of the nutrient value will be available during the first year after you apply the manure, and some will be available the following year(s), and labs often give you estimates of both the first-year nutrient value and total nutrients. Manure is a gift that keeps on giving.

Estimated Nutrient Content of Manure

Below are the nitrogen (N), phosphorus (P_2O_5), and potassium (K_2O) **fertilizer credits** per ton of manure from different livestock types and storage systems. The nitrogen credit compares availability at different times of spreading.

		N				
		TIME TO INCORPORATION				
LIVESTOCK TYPE	STORAGE SYSTEM	> 3 DAYS	1 HR–3 DAYS	< 1 HOUR	P_2O_5	K_2O
Dairy	solid, >20% dry matter	2(1)*	3(1)	3(1)	3	6
Beef	solid	3(1)	4(1)	5 (1)	6	10
Swine	solid	7(2)	9(2)	12(2)	10	8
Chicken	solid	24(5)	27(5)	29(5)	35	26
Horse	solid	2(1)	3(1)	4(1)	5	6

* The first number refers to the first-year nitrogen credit, and the number in parentheses refers to the second-year credit.

Source: "Nutrient Management Fast Facts," University of Wisconsin Nutrient and Pest Management (NPM) Program (see Resources). The actual publication also gives figures for liquid manure and for turkey manure (which is very similar to chicken manure).

This table shows two points very nicely:

1. The faster you can incorporate manure after you spread it, the more nitrogen you'll be able to capture.
2. The nutrient value of chicken manure is much more concentrated than for other kinds of livestock, and it has a much higher phosphate-to-potash ratio than any other type of manure. This is helpful to know if you're amending soils that are very short in phosphorus compared to the amount of potassium they need. If you find yourself in that situation and you plan to purchase manure to take care of the fertility, chicken manure can be a better match than manure from other kinds of animals.

Testing is the best way to know the nutrient value of the manure you're working with, but if you don't have that information you can get some idea of the average value of manure from many good sources. (They vary somewhat, but they're usually fairly close to each other.) One thing to keep in mind is that there's a portion of the manure we apply that plants can use quickly (in the first growing season after we apply it), while some decomposes more slowly and is available later.

Understanding Crop Needs

It's important to understand the nutrient requirements of the crop you intend to raise where you spread manure.

Crops such as corn and potatoes generally need higher nutrient levels (especially for nitrogen) than other crops.

Legumes such as alfalfa, red clover, and birdsfoot trefoil don't need the nitrogen that manure can provide at all if the seed has been properly inoculated and the soil pH is adequate. They rely on the bacteria living in their roots to biologically "fix" nitrogen (see page 22). (When you inoculate the seed you introduce those bacteria into the soil as you plant, and you don't usually need to re-inoculate each year.) Even if they have the bacteria to fix nitrogen, these legumes still benefit from the phosphorus, potassium, and other nutrients in the manure if the soil is deficient in these.

Many leafy vegetables are susceptible to quality problems if they get too much nitrogen, so be careful about using manure on those produce crops.

Whatever crop you're growing, you can get excellent information on the nutrients they need from the university Extension office in your area. Many soil test reports will also provide recommendations for specific crops.

WHERE TO SPREAD MANURE

AFTER YOU HAVE YOUR SOIL TEST results back, you'll be able to see which fields will benefit most from the manure you have available. If all your fields test the same for nutrient levels, maybe you'll just elect to divide the amount of manure you have across all the fields equally — there's nothing wrong with that, as long as applying manure won't cause any of the nutrient levels to become overloaded. It's more typical that some fields need more fertility help than others, and manure is often used as a fertilizer to amend nutrient deficiencies in those fields or pastures.

Manure Provides P and K as well as N

When we focus on using manure for its nitrogen value, we lose track of how good it is at providing other nutrients, especially phosphorus and potassium. **Green manure crops** and old hay fields can provide most of the nitrogen that many farmers need to grow good crops, but sometimes that's not enough. For organic crop farmers who don't have enough old hay fields to plow under to provide the nitrogen they need, manure is usually the cheapest form of supplemental nitrogen, even if they have to buy it. With this awareness, farmers often apply manure first to fields where high nitrogen-demand crops such as corn will be grown. Depending on the amount of nitrogen available from the previous crop and the overall nutrient levels in all the fields, that may or may not be the best place to use the manure.

A WHEELBARROW AND A PITCHFORK work fine for small-scale applications.

Corn is often grown as part of a rotation that includes forage crops with a **legume** component of some sort (alfalfa, clover, or trefoil). Even a poor stand of legume hay or pasture can provide most or all of the nitrogen the corn crop will need. As you consider where to spread manure, look beyond its nitrogen content and capitalize on its phosphorus and potassium value. In most cases where corn follows hay, a very light application of manure will be enough for a good corn crop, and the rest of your manure supply can go to the fields that have the lowest phosphorus and potassium levels.

This is the least expensive way to provide P and K to fields that test low in these. For organic farmers, phosphorus in manure is also the most available form of that nutrient that we can apply.

How Much and How Often?

To use manure as a fertilizer, check your soil test results and the manure analysis to determine how much to apply. Soil tests often provide fertilizer recommendations, so simply consider manure in terms of the equivalent amount of fertilizer it contains.

Let's use potassium as an example. A common potassium fertilizer, potassium chloride, has a potash (K_2O) equivalent of 60 percent, and it has no appreciable nitrogen or phosphorus content, so its fertilizer analysis is listed as 0-0-60. Suppose our soil test recommendations call for applying 150 pounds per acre (lb/a) of K_2O. To determine how much 0-0-60 we should apply per acre:

1. Start with the recommendation (150 lb K_2O/a) and divide it by the decimal equivalent of the percent potash in the material you want to use. Potassium chloride, 0-0-60, is 60 percent potash. For this calculation use its decimal equivalent of 0.6.
2. Plug these numbers into the formula.
3. The result: 150 lb/a divided by 0.6 = 250 lb/a of 0-0-60.

The same method works for phosphorus, using the phosphate numbers from the soil test recommendations.

In a similar way we can determine how much manure to apply to meet the soil test recommendations. The "textbook" approach to figuring out how much manure to apply is just a matter of knowing how many nutrients are in the manure you have and how many nutrients you need to apply to the field. Then it should be simple to select the right application rate and put it on the land, right?

Well, maybe it's not quite that simple. The problem is that the proportions of the various nutrients in manure are fixed, and the proportions we need to apply to our fields vary from place to place. Soil tests almost never show that the nutrients you need to apply are in the same proportions that are in the manure, so you have to decide how best to employ this resource. Usually we pick one or two of the most pressing needs and choose a rate of manure that will meet (or come close to meeting) those needs. This often means that some other nutrients will be either under- or over-applied.

Here's an example. A soil test reports that a field is slightly low in phosphorus and very low in potassium. The recommendation from the lab calls for applying 50 lb/a of phosphate and 300 lb/a of potash. Suppose we have horse manure on hand. According to the manure nutrient tables above, average horse manure will supply 5 lb of P_2O_5 and 6 lb of K_2O per ton. To meet the phosphorus recommendation, we could apply 10 ton/a of the horse manure. But that amount of manure would provide much less potassium than the soil test recommended.

If you apply the manure at the 10 ton/a rate, you could just apply another form of potassium (potassium chloride or potassium sulfate) in addition to the manure to make up the difference. Or you can increase the rate of manure to 50 tons per acre to meet the potassium recommendation.

If you do that, however, you'll apply a lot more phosphorus than recommended. Fifty tons of manure per acre is a very high application rate. If the field doesn't already have excessively high levels of phosphorus, this might be acceptable. If the level in the field is already high, applying this much manure could drive the phosphorus up to where it could cause environmental or agronomic problems.

There are advantages and drawbacks to either way of using the manure, depending on the background fertility and whether you have extra manure you can apply or extra money you can spend on fertilizer.

SMALL GROUND-DRIVEN SPREADERS can be pulled behind small tractors, ATVs, or horses.

A LARGE LIQUID MANURE SPREADER with injectors.

When What You Have Isn't As Much As You Need

Sometimes there isn't enough manure available to completely meet the nutrient needs of the land you're working with. Here are a couple of principles to keep in mind.

The fields with the lowest fertility will respond best to any nutrients you're able to add. If you have a limited supply of manure, you should apply manure to the least fertile fields first.

The first unit of fertilizer you apply to a low-testing soil will give you the biggest response. For instance, imagine a situation where a soil test indicates that you should apply 100 lb/a of phosphate. You should get a profitable response from applying the entire 100 lb/a, but if you aren't able to do that, the first 50 lb/a that you apply will give you a greater response than the second

A TYPICAL REAR-DISCHARGE SOLID MANURE SPREADER behind a tractor.

50 lb/a increment will. The yield response you get from manure applications works this same way — the first increment you apply gives a disproportionately high response compared to the successive increments.

If the soil nutrient levels are low, whatever you do apply will help. If all the fields have roughly the same need for extra nutrients, then put the same rate of manure on all those fields, even if it means not putting as much on each field as you'd like.

A Few Other Things to Keep in Mind

Here are some miscellaneous additional thoughts about applying manure.

Composted manure. If the manure you're going to use has been composted (or even just well rotted from being aged a long time) it won't behave exactly the same as fresh manure. Aside from nitrogen, most of the nutrients present in the original material are preserved during composting, but the volume is reduced substantially. This means that most of the nutrients in composted manure are more concentrated than in the fresh material. Accordingly, we can use much lower rates of compost to improve fertility than if we were using fresh manure. Some of the nitrogen content of the original manure will be lost during the composting process, so this element won't be concentrated to the same degree that the others are.

Forage-crop nutrient removal by mechanical harvest. Each ton of forage dry matter you harvest from a field removes 12 to 15 pounds of P_2O_5 and 50 to 60 pounds of K_2O. Depending on the type of soil you have and its starting fertility, you may have to replace these nutrients in order to keep the land productive. Where the starting fertility is low, you may need to supply even more than these amounts of nutrients as you harvest forage in order to get the land to its optimal productivity. On the other hand, if you start with excessively high levels of nutrients, you can also use forage harvests to bring these nutrient levels down to the more optimum ranges.

Grazing management. When livestock graze, they don't just eat — they also poop and pee. Since most of the nutrients animals eat pass through their bodies, with good grazing management we can recycle most of what the animals are eating from the pasture. This means controlled grazing (such as **rotational grazing** or management-intensive grazing), not just having a pasture area where livestock can roam around at will for as long as they want.

If you follow excellent grazing practices, up to 60 percent of the nutrients that dairy cattle harvest can be returned to the pasture (for beef, it could be 75 percent). If you have ruminant livestock and graze them, I encourage you to develop your

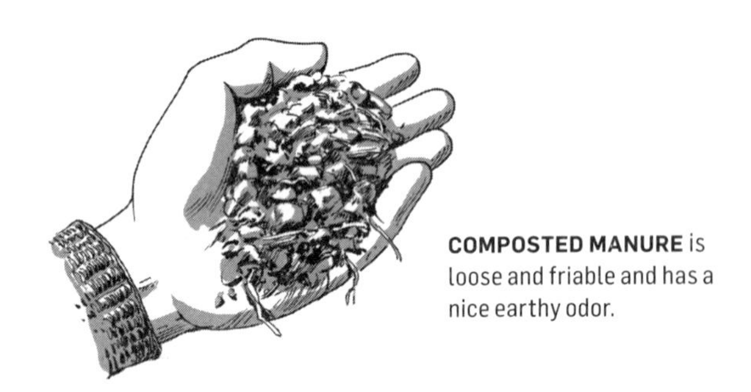

COMPOSTED MANURE is loose and friable and has a nice earthy odor.

grazing system according to the principles of rotational grazing. For an excellent introduction to the principles of rotational grazing, see "Pastures for Profit," published by the University of Wisconsin-Extension (see Resources).

WHEN TO SPREAD MANURE

THE SEASON YOU APPLY MANURE affects how it performs as a fertilizer, which influences the yield you get, and it may also affect the quality of the crop. The best time to apply manure to a certain area depends on the soil in that area and the crop you intend to grow. Here are some general rules of thumb.

Perennial Forage Crops

If the crop is an existing perennial forage stand (a hay field or pasture), I recommend applying the manure in the late summer or fall. During that time of year, perennial plants are designed to take up nutrients and energy and store them in their roots and **crowns** (the base of the stem, located at the surface of the soil) so they can live through the winter and send up new shoots in the spring. *Applying manure at this time of the year allows these plants to take up whatever portion of the nutrients is in the soluble form.*

This not only strengthens the plants, it conserves the nutrients by holding them so they won't leach away during the winter. This is especially important with nitrogen and potassium, the two nutrients that are most likely to be present in soluble form when the manure goes on the field.

Why fall instead of spring? In the spring, the plant is bringing nutrients up out of the roots and crowns and putting it into the new growth, so the flow of nutrients is the opposite of what happens in fall. Nitrogen and potassium that are applied in the spring get taken up readily by the plants and much of that goes right into the herbage of the first crop of forage. This means that the fertilizer value of those nutrients behaves more like a "flash in the pan," and there isn't a lot of residual benefit for the successive cuttings or grazings of the forage.

Not only that, but if the forage is for dairy cattle, too much potassium early in the year can cause excessive levels of that nutrient in the feed for the cows, which can lead to metabolic disorders like **grass tetany** (hypomagnesemia) or **milk fever**. *Nutrients applied to perennial crops in the fall strengthen the vigor of the entire plant,* and they can be used by the plant during more of the following season.

MANURE AND GRAZING LIVESTOCK

If you decide to spread manure on a pasture to bump up the fertility, be aware that it will be quite a while before the livestock graze those pastures again (at least if the manure is from the same species). All livestock come designed with a self-protection feature that warns them against grazing too close to their own feces, because doing so would likely expose them to a fresh batch of internal parasites (worms), shed as eggs through the animal's manure. (Urine doesn't contain worm eggs and presents nowhere near the refusal problem that dung patches do.) Therefore, when we spread fresh manure on a pasture, we effectively coat the entire grazing area with material that warns away the animals we're trying to feed.

There are two ways to get around this problem. One is to spread the manure in the fall when the stock are done grazing (by the time spring rolls around, there is unlikely to be as much refusal). The other is to spread composted (or even very well-rotted) manure. This material doesn't smell like manure either to us or to the livestock, and it won't cause feed refusal. Because the nutrients in compost or well-rotted manure are in a very stable form (and any internal parasites would have died by then), it's okay to spread this material whenever it's convenient.

Cover Crops

If you use **cover crops** in your cropping system (a very good practice!), apply manure just before you plant the crop to conserve nutrients and beef up the benefits of the cover crops. This is true for whatever season you plant them.

Annual Row Crops

Many farmers like to apply manure in the fall of the year before they plant corn, soybeans, or spring-seeded small grains. It's a way to cut down on some of the work that needs to be done in the spring and often helps control perennial weeds that have established in the preceding crop. This can be a good practice in some cases, but it's best to do it in conjunction with a cover crop or an existing forage stand if at all possible, so that you won't lose nutrients to leaching over the winter.

Never apply manure to bare ground in the fall on fields with sandy soil, since many of the soluble nutrients will most likely leach away over winter. On sandy ground, it's almost always better to apply the manure in the season you'll plant, so that the growing plants can use the nutrients as they become available.

Organic Vegetable Crops

If you grow vegetable crops, incorporate manure into the ground at least 90 days before harvesting any crop. If the crop you are growing has direct contact with the soil surface (potatoes, onion, radishes, etc.) then you must wait at least 120 days between application and harvest. This time restriction doesn't apply if the manure is composted in accordance with the NOP rules. Aged manure that has not gone through the documented composting process, even if it's as old as the hills, is still considered to be raw manure, and you need to follow the time interval rules.

CALIBRATING MANURE SPREADERS

Knowing how much manure we want to spread is important, but you must also know the amount you are actually applying, so as to adjust it as needed. If you use a spreader, it's a good idea to calibrate it so that you know what your application rate is.

Calibrating a spreader boils down to measuring the weight or volume of manure in one spreader load and then calculating the area that one load covers. If we divide the weight (usually expressed in tons) or volume (usually expressed in gallons) by the area of land (usually expressed in acres) we have our application rate. Knowing the nutrient value of the manure we spread allows us to calculate how many pounds of the major nutrients we're applying.

Many soil and water land conservation departments or Extension offices offer help with calibrating manure spreaders. If that isn't an option or you prefer to do this yourself, it's not hard. There are a number of good references available that give detailed information on calibrating manure spreaders. One recommended resource is the University of Vermont (see Resources).

There are several ways to calibrate manure application systems.

METHOD 1: CALCULATING BY FULL VOLUME

The first method involves doing a simple calculation considering the volume (or weight) of the full spreader and the amount of land covered by that load. You'll determine the amount of weight your spreader can hold in tons, and divide that by the amount of land in acres covered by one full load.

Step 1: Determine manure quantity. If you have liquid manure, the specifications for your spreader will give you an accurate figure for the volume of a full spreader load. With solid manure, the most accurate way to know how much weight you apply with each load is to weigh the full spreader. University Extension offices or land and water conservation departments in your area may have portable scales you can use to do this. Otherwise it will involve a couple of trips across a load scale — one with the spreader empty and another with it full. These types of scales are common at feed mills and trucking companies.

Another method for solid manure is to estimate the volume of the load you have, and then use average manure density values to estimate the weight of a load. According to the University of Vermont, solid cattle manure varies from 55 to 62 lb/cubic foot, depending on moisture and the type of bedding used. Not only does the density of solid manure vary a lot, but it's also hard to estimate just how much volume is in a load because of the many different ways people load spreaders. This method is a lot better than just guessing, but it would be easy to make a substantial miscalculation this way.

Step 2: Determine land covered. Next, measure the amount of land you cover with one load. For example, let's say we find that the spreader covers a swath 20 feet wide and we can drive 435 feet to empty the load. This covers 8,700 square feet (20 feet times 435 feet). We need to convert this to the equivalent land area expressed as acres. There are 43,560 square feet in one acre. The land area we covered is 8,700 square feet, so if we divide this by 43,560 square feet per acre, we get about 0.2 acre.

Step 3: Calculate rate. Finally, divide the weight of the manure in a full load by the area covered. For instance, if we know that our spreader holds 6 tons of manure with each load and we cover 0.2 acres, our application rate is 6 tons divided by 0.2 acres, or 30 tons per acre (t/a). That's a fairly heavy application rate. In most cases it's probably not a good idea to spread heavier than that and 20 t/a would be even better for a single application. (With poultry manure, you should use much lower application rates than you would with larger livestock because the nutrient content is much higher. I usually suggest applying no more than around 2 t/a of poultry manure.)

METHOD 2: CALIBRATING BY TARP

If you can't weigh the spreader and you aren't confident of the volume of the spreader, there's another simple way to calibrate. This method involves using three (or more) tarps (or uniform-sized pieces of heavy plastic sheeting) that you put in the path you're spreading. With this method, you collect the manure that lands on each of the sheets and actually weigh it. Suppose the sheets you use are 6 feet by 6 feet. That's 36 square feet per sheet. Now let's say the average weight of the manure we caught on the sheets is 20 lb. To find the equivalent application rate in tons per acre, we can use this formula:

- Average weight of the manure on the sheets (in lb) divided by the area of each sheet (in square feet) = lbs of manure per square foot.
- Next, take the result from the first step and multiply it by 43,560 (the number of square feet in one acre).
- Finally, divide the number resulting from the first two steps by 2,000 (the number of pounds in a ton).

Plugging in the numbers from the example, this is (20 lb/36 square feet × 43,560 square feet/acre) divided by 2,000 lb/t = 12 t/a (rounded to the nearest ton).

The reason for using three sheets is that most manure spreaders don't do a very good job of macerating the manure as it comes out, and there can be a lot of variability in what actually lands on the ground from one place to another. You can also use these three tarps to see how the application rate varies from one side of the spreading path to the other. A lot of beater-type spreaders put more manure directly behind than they do on the sides of the travel path. To check this, place one tarp in the center of the path and one on each side of the spreading pattern. If you see a much heavier application rate in the center of the spread, you may choose to overlap your spreads a bit to compensate.

Adjusting Spreading Rate

If you find that your spreading rate is higher or lower than you'd prefer, change the rate by one of three methods: adjust your spreader, adjust your speed, or adjust your load. You may or may not be able to easily adjust the rate that you're spreading. Some spreaders allow you to change settings to adjust the rate somewhat. If the terrain is forgiving, you may be able to adjust your travel speed. Driving slower (by driving in a lower gear) will increase your application rate, and driving faster while you spread will decrease your rate. With a box-type (horizontal beater) spreader and solid manure, adjusting the depth of the manure in the load may give you some flexibility in the rate you apply. Loading the spreader to a lower height will result in a lower application rate if you can't increase the ground speed enough, but it will also mean more trips across the field.

Liquid Systems

This discussion has focused on solid manure systems, because they are more common on small farms. If you use a liquid manure system, calibrating your spreader is even easier, because the manufacturer's information for your spreader will tell you its capacity in gallons. Unlike a box-type solid manure spreader, when a liquid spreader is full, it's full — there's no guesswork involved. Calibration becomes a simple matter of dividing the number of gallons in one full load by the area of land you cover as you spread it. Simply multiply the width of the spread by the length in feet, and divide this number of square feet by 43,560 to determine the acreage. Then divide the weight of the manure by the acreage to determine the application rate.

SMALL-SCALE APPLICATIONS

IF YOU'RE A GARDENER OR a small-scale produce grower, you may be interested in spreading manure by hand. The same principles apply for small plots: determine the amount of manure you'd like to apply, and then distribute it uniformly over the appropriate amount of land area. Depending on the lab you use for soil testing and the needs of your crop, you may get recommendations based on pounds of nutrients per 100 square feet or per 1,000 square feet. Weigh the container you use to haul the manure, such as a 5-gallon bucket, and determine how many of those buckets you need to apply over the desired area. Keep in mind that this isn't as precise as brain surgery — it's perfectly fine to round numbers off to make things easier.

Here's an example: You have a garden that's 35 feet wide and 70 feet long. That's 2,450 square feet. The soil test report recommends applying various nutrients, but phosphorus seems to be the most limiting factor. The report calls for applying 4.5 lb of P_2O_5 per 1,000 square feet.

1. To find out how many pounds of phosphate we should apply to our garden, start by multiplying the fertilizer recommendation (in lb/1,000 square feet) by the area of the garden (in square feet) and divide by 1,000: 4.5 lb $P_2O_5 \times 2{,}450/1{,}000 = 11$ lb P_2O_5.
2. Decide what kind of manure you'll use to meet that need, and then calculate the number of pounds of that manure you'll need to apply to do the job. Poultry

manure is a great source of phosphorus, and based on average values it should furnish 35 lb P_2O_5 per ton. Divide 35 by 2,000 to find out how many pounds of P_2O_5 there are in a pound of poultry manure (35/2,000 = 0.0175 lb P_2O_5/lb of manure).

3. We already determined you want to apply 11 lb of P_2O_5 to the garden area, and you know that each pound of poultry manure should give you around 0.0175 lb of P_2O_5. If you divide the desired amount of phosphate (11 lb) by the phosphate content of the manure (0.0175 lb P_2O_5/lb of poultry manure) you will know how many pounds of manure to apply to the garden: 11 lb P_2O_5 divided by 0.0175 lb P_2O_5 per pound of manure = 630 lb of poultry manure. (Again, these numbers are rounded off. It's okay to follow the same rules for manure math that you would when you play horseshoes: being close is good enough.)

WEIGH PLASTIC PAILS full of manure to determine application rates for small areas.

YOU CAN SPREAD composted manure with a feed bag right in the rows of your garden.

TOP DRESSING OR INCORPORATING?

THE WAY YOU APPLY MANURE has a lot to do with how much benefit your land and crops get from it. Manure can either be spread on top of the ground (**top-dressed**) and left for weather and the soil biology to deal with, or it can be **incorporated** into the soil by cultivation.

Whatever system you use, you should be putting the manure where it does the most good. You should also apply it at rates that are appropriate for what the soil needs with respect to the crops you'll grow on that piece of ground. An application system that can distribute the manure uniformly across the area being spread is always the best.

Top-dressing manure is often the quickest, easiest, and least expensive way to apply it. Top-dressed manure can act like a mulch, shading the ground from the hot sun and helping to conserve moisture. There are also some drawbacks to doing this. Some of the nitrogen in uncomposted top-dressed manure will be lost as ammonia volatizes, and less soluble nutrients in manure on top of the ground will take a long time to become available to the crop. Manure on the surface of the soil can be washed away during intense rainfall or rapid snowmelt. During dry weather, some of the manure on the soil surface may not continue to break down until more rainfall wets it again.

Incorporating manure means putting it underground either by tilling it in after spreading it or by injecting it (for liquid manure systems). This requires more equipment, and it usually is only practical where the ground is going to be worked up to prepare for the next crop. Incorporating manure cuts down on odor and conserves more of the nitrogen in the manure, because any ammonia gas that would otherwise escape to the atmosphere is held in the soil.

If you use ordinary tillage implements to incorporate manure, it's usually better to work the manure into the ground with a disc or harrow rather than using a moldboard plow. Organic matter like manure, or even a heavy sod that is only plowed under, often gets buried at the base of the plow layer where there is little oxygen. In this condition, the manure will basically be entombed in an anaerobic state. Without enough oxygen, the manure doesn't decompose naturally as it would if air were available. Anaerobic decomposition proceeds very slowly and can actually produce gases that are toxic to the desirable organisms in the soil.

CHAPTER FOUR

RULES, REGULATIONS, AND MARKETING

Life just isn't as simple as it used to be. Nowadays there are even laws about how we can spread manure! There may be regions where even more regulations apply, but here are a few common ones to keep in mind.

NUTRIENT MANAGEMENT PLANNING

MANY AREAS OF THE UNITED STATES are covered by nutrient management planning rules that govern where, when, and how much manure can be applied to cropland. These rules came about because of widespread groundwater and surface water pollution that arose from excessive fertilization (with both manure and other fertilizers) and poor cropping practices. This combination has allowed many drinking water supplies to become contaminated with nitrate and has caused both nitrate and phosphate pollution in surface waters (lakes, streams, rivers, and even parts of the ocean). The specifics of nutrient management planning regulations vary from place to place, so it's a good idea to familiarize yourself with whatever rules govern your area.

In general, the rules are designed to prevent fields from receiving too much nitrogen and phosphorus, so soil testing is usually part of the program. The amount of manure you can apply to a field will probably be limited, considering the existing levels of nutrients in the soil and the requirements of the crop. In addition, the rules often have guidelines designed to minimize the chance for losing nutrients in manure to runoff. This may mean limiting manure applications to times of the year when the ground isn't frozen, limiting the slopes to which manure can be applied, and limiting how close manure can be spread to wells, waterways, and ditches.

USING MANURE IN ORGANIC FARMING SYSTEMS

For a farmer to be able to represent his or her products as organic, the production system has to follow all of the requirements of the U.S. Department of Agriculture's National Organic Program (NOP). This isn't just a requirement for larger, commercial farms — it applies to everyone who sells products that are represented as being organic. Even small-scale organic operations that are exempt from certification still have to follow all the NOP rules for organic production if they market their products as organic.

The organic rule prohibits almost all synthetic fertilizers and pesticides from being used in organic production systems, but manure, being a natural product, is permitted. The rules allow even manure from conventional farms to be used as a soil amendment for organic crop production, as long as no prohibited materials (such as insecticides) are added to it. If there's bedding with the manure, it can't contain prohibited materials like sawdust or chips from treated wood.

This brings up a question that many organic farmers are wrestling with. Conventionally raised livestock are often treated with hormones and antibiotics to increase their growth rate or milk production or to synchronize breeding. Residues from these treatments can pass through the animals into their manure. In addition, pesticide residues from livestock feed may contaminate their manure. More recently, people have started to question the effect of feed from crops that have been bred to withstand many popular herbicides, especially **glyphosate** (Roundup). Many organic farmers have chosen not to use conventionally produced livestock manure, even though it's legal, because of these concerns.

DEALING IN USED FEED: SELLING MANURE

IF YOU HAVE AN AMPLE SUPPLY of manure and would like to sell some, there are a few things to keep in mind.

Legal Considerations

Manure sales aren't regulated along the same lines as food products, but there still may be some legal requirements you need to be aware of to stay on good terms with the law. Local zoning regulations may govern what kind of commerce goes on in your neighborhood, traffic and parking, signage, and other associated issues. Check with your local authorities to find out if zoning affects your business.

Nutrient Analysis

It's important to be able to give an accurate representation of your product. Follow the nutrient testing guidelines I discussed in the previous chapter to enable you to describe your product to potential buyers.

PACKAGING MANURE FOR SALE in small quantities doesn't necessarily require fancy equipment.

Logistics

What quantities of manure are you interested in selling, and what are the units? By the bag, by the wheelbarrow, by the pickup load, or by weight? If you sell by volume, you should have some idea how much a given volume of manure weighs, so you can get a fair price for its nutrient value. If you sell by weight, you'll need to have access to a scale. What about loading? Will you be forking the manure into someone's container, or using a tractor and loader, or will the customer be responsible for this?

Marketing

How and where will you advertise the manure? Try classified ads in farm-related publications or even your local newspapers. Web-based outlets like craigslist can give you lots of coverage for free, and the customers self-select their shopping interests. Post an ad on the bulletin boards in farm stores, feed mills, or garden supply businesses in your area. A brief description of your product ("stacked horse manure for sale") is enough information for a classified ad or bulletin-board poster, but be prepared with a detailed description of the product and related information for follow-up inquiries.

Pricing

There are several ways to estimate how much manure is worth. In the end, it boils down to how much someone is willing to pay for it. One way to come up with a starting figure is to put a value on the nutrient content of the manure and compare it to how much those nutrients would cost in the form of fertilizer. This isn't a perfect method, because even though most of the monetary value of manure is in its macronutrient content, manure is also a great source of minor and micronutrients. It's also very difficult to put a monetary value on the effects of manure's organic matter and other biological contributions. Chemical fertilizers have their place, but it's impossible for them to provide these other benefits that manure brings.

As a starting point, we can assign manure a conservative monetary value by considering how much nitrogen, phosphorus, and potassium it has per ton. The content of these nutrients and how much it would cost to buy them in commercial fertilizer gives us an objective guideline.

Here's an example.

Nutrient value. Let's say we have beef manure available, and we assume its nutrient value is close to the averages presented in chapter 3. Conservatively each ton of manure will provide 3 lb of nitrogen, 6 lb of phosphate (P_2O_5), and 10 lb of potash (K_2O) in the season after we apply it.

Current price. A call to the local fertilizer dealer gives us the current prices of several common forms of these nutrients: urea (45-0-0), diammonium phosphate (18-46-0), and potassium chloride (0-0-60). We find out that urea is selling for $500 per ton, diammonium phosphate is $550 per ton, and potassium chloride is $800 per ton.

Price per ton. We can calculate the price per ton of the nitrogen, phosphate, and potash from these sources easily. Here's how: We'll start with the diammonium phosphate, because it contains both phosphate and nitrogen, and if we were to use this product we'd need to buy less of another form of nitrogen. The price per pound of the entire product is $550 divided by 2,000 lb (one ton), or $0.275/lb. However, only 46 percent of each pound of this fertilizer is actually phosphate, so we need to divide that price by 0.46: $0.275/lb divided by 0.46 = $0.60/lb of actual phosphate.

Comparison. If there are 6 lb of phosphate in a ton of manure, the phosphorus value is 6 × $0.60 or $3.60. This is the same amount of phosphate we could have gotten from 13 lb of diammonium phosphate (6 lb of actual phosphate divided by the phosphate content of diammonium phosphate, 0.46, equals 13 lb).

Estimating nitrogen. Now, if we had bought the 13 lb of diammonium phosphate, we also would have gotten 2.3 lb of nitrogen, because that product is 18 percent N (13 lb × 0.18 = 2.3 lb). This is most of the amount of the 3 lb of readily available nitrogen that the 1 ton of manure would have supplied. If we substitute commercial fertilizer for manure, we'd need to account for an additional 0.7 lb of N that we would have gotten from the ton of manure. Urea is a common nitrogen fertilizer, and the dealer tells us it costs $500/ton. Using the same approach we used for phosphorus, we go through this sequence: $500/2,000 lb = $0.25/lb of urea. Since urea is 45 percent N (45-0-0), we divide the price per pound of urea by the percent N and we find that nitrogen from urea costs: $0.25/lb divided by 0.45 = $0.56/lb of N. If we multiply the price per pound of N by the amount of N we want to provide from the urea, we get 0.7 lb N × $0.56/lb N, or $0.39.

Estimating potassium. The last of the "Big 3" nutrients that give manure most of its direct fertilizer value is potassium. The most common form of potassium fertilizer is potassium chloride, and when we checked the price it was $800/ton. Applying the same method we used in the examples above: $800/2,000 lb = $0.40/lb of potassium chloride. At 60 percent K_2O, this equals $0.40 divided by 0.6 or $0.67/lb of potash. From the table, average solid beef manure provides 10 lb of readily available potash/ton, so the potash value is $0.67/lb of potash times 10 lb of potash per ton of manure, or $6.70.

Total it. Adding up the values of the equivalent amounts of these commercial fertilizers that 1 ton of average beef manure would provide, we get $3.60 (phosphate) plus $0.39 (nitrogen) plus $6.70 (potash), or $10.69/ton of manure.

As I mentioned above, this is a very conservative value. Manure provides some amounts of all the other nutrients required by plants; some of the total nutrients in manure are released after the first year and aren't included in the credits listed in the table, and the value of the organic matter and the biological benefit of manure are very difficult to quantify.

If you're an organic farmer, the value of manure is even greater because the selection of organic fertilizer sources to substitute for the nutrients in manure is limited. Even when we can substitute organic fertilizers for similar conventional products they are often more expensive than the conventional ones.

BUYING MANURE

If you're buying manure, you can use most of the information in the preceding section to help you determine how much you can afford to pay. It's true that spreading manure is usually more work than using ordinary fertilizers, but its benefits outweigh this concern. Nevertheless, logistics are important. You'll have to be able to haul, handle, store, and apply the manure appropriately for whatever farming or gardening system you use. Whether the manure is fresh or it's been composted or rotted affects how easy it is to work with and how stable it is in the soil.

It's also important to remember that not all manure is the same. For instance, poultry manure generally has more concentrated nutrient levels than cattle manure. If there are several types of manure available at different prices, understanding how to calculate the nutrient value of each will help you determine how much one type is worth compared to another.

Whether you're buying or selling manure, the real monetary value still ends up being what the buyer is willing to pay. The information in this section should help you determine how much you can afford to pay when you need to make these decisions.

CONCLUSION

I hope you find the information in this book useful and interesting. Understanding more about manure can help us to be better farmers, gardeners, and stewards of the earth. There is way too much waste in the world, and understanding more about the cycle of life and how we can fit into it can help us make better use of the resources we have.

I see manure as not only interesting, but amazing and glorious as well. I hope you share a little of that with me. Have fun spreading the wealth!

GLOSSARY

Actinomycetes. A type of soil bacteria that grow in rootlike colonies that resemble fungal strands (hyphae). They are good at breaking down resistant organic materials like cellulose, give soils their characteristic earthy odor, and are the source of many antibiotics used in medicine.

Aggregate. A discreet package of soil that is formed of the individual particles of sand, silt, clay, and organic matter. When a soil is well aggregated, there are channels or pore spaces between the aggregates that allow water, air, and soil organisms to move more freely through the soil.

Amendment. A material that is added to the soil to improve its fertility, reaction (acidity/alkalinity), or physical characteristics.

Anthelmintics. Naturally occurring or synthetic materials that inhibit or kill internal parasites.

Bedded pack. A system in which animals are fed and bedded in the same area for an extended time so that a mound of bedding, manure, and waste feed accumulates.

Beneficial elements. Chemical elements that may be required by some, but not all, plants, or that help plants to grow better, even if they aren't absolutely necessary for plants to survive.

Biological nitrogen fixation. The process of converting atmospheric N into a compound that plants can take up and utilize.

Carbonaceous ("brown") material. A type of organic matter with a very high carbon content in relation to the amount of nitrogen it contains. Examples: sawdust, straw, corn stalks, rice hulls, and wood chips.

Compost barns. Livestock housing structures that use a layer of carbonaceous bedding to absorb manure, and that require daily mechanical mixing of the bedding to actively compost the material while it's still being used as bedding.

Composting. An aerobic decomposition process that usually implies frequent mixing/turning to generate high temperatures.

Cover crops. Crops that are grown exclusively or primarily as a source of fresh organic matter for the soil rather than to be harvested as a crop for human food or livestock feed.

Crown. The part of a plant where the stem and roots merge.

Daily haul system. A manure handling system that is accomplished by collecting and hauling each day's manure production out to a field or pasture for spreading.

Essential elements. Chemical elements that are required by all plants for their survival.

Fertilizer credit. Credit for various plant nutrients that can be supplied by materials other than fertilizers.

Glyphosate. A specific formulation of nonselective, synthetic, systemic herbicide that is widely used in conventional agriculture and horticulture.

Grass tetany (hypomagnesemia). A metabolic disorder of cattle characterized by low levels of magnesium in the blood. It is caused by the animal consuming forages that have very high potassium and/or very low magnesium levels. It is also called grass staggers.

Green manure crops. Leguminous cover crops that are grown to provide high amounts of nitrogen to crops that will be grown next in a rotation.

Gypsum. Hydrated calcium sulfate. Often used as a soil amendment, especially in organic farming systems, as a source of sulfur and calcium and to improve soil aggregation/tilth.

Humanure. Human excrement.

Humus. The type of organic matter that is left in the soil after all soil organisms have digested the usable fractions of the material. It has very small particle size and is resistant to further breakdown in the soil. It helps bind soil aggegates together and enhances the ability of the soil to hold onto and exchange plant nutrients.

Hydrogen sulfide gas. (H_2S). A poisonous gas that is released from organic matter that decays under anaerobic conditions.

Incorporate. To till or mix into the soil.

Ions. Atoms or molecules that carry an electrical charge from having an excess or deficiency of electrons.

Legumes. Plants in the pea or bean family. Other examples include alfalfa, clovers, vetches, lupines, and trees like locusts and acacias.

Macronutrients. Plant nutrients not supplied directly from air or water that plants need in very large amounts. Examples: nitrogen, phosphorus, and potassium.

Manure. The dung (feces) and urine, mostly of farm animals, with or without other materials used as bedding.

Micronutrients. Chemical elements that are required by plants in very small amounts, such as zinc, copper, boron, manganese, iron, chlorine, nickel, and molybdenum.

Milk fever. A metabolic disorder of livestock caused by a deficiency of calcium in the blood.

National Organic Program (NOP). A set of rules governing the legal requirements of organic agriculture in the United States; administered by the U.S. Department of Agriculture.

Night soil. A traditional term for human excrement applied to agricultural lands.

Nitrogenous "green" material. Organic matter characterized by a relatively high proportion of nitrogen in relation to the amount of carbon it contains. Examples include fresh grass clippings or high-quality hay, alfalfa pellets, soybean meal, and most manures.

Peds. The scientific term for soil aggregates.

Phosphate (P_2O_5). The oxide form of phosphorus that is traditionally used in fertilizer analysis and fertility recommendations.

Pigerator. A system that uses hogs to aerate and loosen a bedded pack manure storage area.

Potash (K_2O). The oxide form of potassium that is traditionally used in fertilizer analysis and fertility recommendations.

Rhyzobial bacteria. A type of bacteria that are associated with most leguminous plants. They contain an enzyme called nitrogenase that allows them to convert inert nitrogen gas from the atmosphere into the forms used in their bodies. They also provide plant-available nitrogen to their host plants and other organisms living in that plant community.

Rock phosphate. A mined rock product that contains relatively high levels of phosphorus; commonly used as a phosphorus fertilizer in organic crop production.

Rotational grazing. A controlled grazing system that relies on grazing cattle in discreet paddocks for a short time. After that forage is grazed down to the appropriate height, the cattle are moved to the next paddock, and they aren't allowed to regraze the area until it's rested enough for the plants to be at their prime condition again.

Slurry. A semisolid mix of manure, bedding, and water.

Static pile. A manure pile that is undisturbed during storage; it decomposes anaerobically over a long time.

Top-dress. To apply fertilizer or manure to the surface of the soil without incorporating it.

Vermicomposting. A system of composting manure that relies on the activity of earthworms managed specifically for that purpose.

ACKNOWLEDGMENTS

I've spent much of my life as a student of the earth, especially soils, agronomy, and farming systems. Very little of what I know is original discovery. Along the line, I've had many excellent teachers, including my father, Harvey Kopecky, who had a lifetime of farming experience and passed some of that along to me. My brothers Howard and Tom continued their farming heritage by becoming high school agriculture teachers and taught me many things. During my formal education, the person who had the greatest influence on my thinking and learning is my friend and mentor, Dr. Larry Meyers. Farmers, colleagues, and farming and gardening friends whose names would fill this little book have taught me more than I remember, and I'm grateful to them for all that they have shared with me.

Above all, I thank God for the amazing creation he allows us to live in, and especially for his unending love and mercy that invites us all to eternal life through his son, Jesus Christ.

RESOURCES

REFERENCES

Brady, Nyle C., and Ray R. Weil. *The Nature and Properties of Soils,* 14th ed. Prentice Hall, 2007.

Cooperband, Leslie. "The Art and Science of Composting." Center for Integrated Agricultural Systems, 2002. (Available online at: www.cias.wisc.edu/wp-content/uploads/2008/07/artofcompost.pdf)

Hernandez, Jose A. and Michael A. Schmitt. "Manure Management in Minnesota, rev. ed." Bulletin 03553. University of Minnesota Extension, 2012.

Jenkins, Joseph. *The Humanure Handbook,* 3rd ed. 2005. http://humanurehandbook.com.

Joern, Brad C., and Sarah L. Brichford. "Calculating Manure and Manure Nutrient Application Rates." Bulletin AY-277. Purdue University, 1993.

Jokela, Bill. "Manure Spreader Calibration." University of Vermont-Extension. http://pss.uvm.edu/vtcrops/articles/ManureCalibration.pdf.

Kirsten, Amanda, ed. *The Penn State Agronomy Guide 2013–2014.* College of Agricultural Sciences, Penn State University Extension, 2013.

Magdoff, Fred and Harold van Es. *Building Soils for Better Crops,* 3rd ed. Handbook series book 10. Sustainable Agriculture Research and Education, 2009.

Minnesota Department of Agriculture. Nutrient/Manure Management Information Tables. University of Minnesota-Extension.

Morse, Debra Elias. "Composting Animal Mortalities." Minnesota Department of Agriculture, 2009. www.mda.state.mn.us/news/publications/animals/compostguide.pdf.

Nutrient and Pest Management Program. "Nutrient Management Fast Facts." University of Wisconsin, 2013.

Schulte, E. E., L. M. Walsh, K. A. Kelling, L. G. Bundy, W. L. Bland, R. P. Wolkowski, J. B. Peters, and S. J. Sturgul. *Management of Wisconsin Soils*, 5th ed. Bulletin A3588. University of Wisconsin-Extension, 2005.

Thompson, Louis M. *Soils and Soil Fertility*. McGraw-Hill, 1952.

Undersander, Dan, Beth Albert, Dennis Cosgrove, Dennis Johnson, and Paul Peterson. "Pastures for Profit: A Guide to Rotational Grazing." A3529. University of Wisconsin-Extension, 2002. www.learningstore.uwex.edu/assets/pdfs/A3529.pdf.

Van Slyke, Lucius L. *Fertilizers and Crop Production*. Orange Judd, 1949.

FURTHER READING SUGGESTIONS

There are lots of great historical and scientific references available that deal with manure as a component of soil management, and I encourage you to seek these out both for enjoyment and to gain a deeper understanding of manure management than I have space for here.

Practical students of soil science will do well to get a copy of the book *Building Soils for Better Crops* by Fred Magdoff and Harold van Es. This is a book that every farmer and gardener should have — it's easy to read, but it has plenty of scientific background. It will help you understand both the how and the why of soil management. For those who want a bit more technical information, the introductory college soils textbook *The Nature and Properties of Soils* by Nyle C. Brady is my favorite.

Some older soils books have valuable information on manure management as well, especially for those who are interested in organic or other sustainable methods of agriculture. A couple of nice older books with excellent information on manure management are *Fertilizers and Crop Production* by Lucius L. Van Slyke and *Soils and Soil Fertility* by Louis M. Thompson. These books were published in the mid-1900s, but much of the information is as helpful and practical today as it was then.

Universities all over the nation (and the world, for that matter) have published useful reference books and websites, and you should consult these for information that is tailored to the region where you live. One that I have drawn from heavily for this book is the University of Wisconsin-Extension Bulletin A3588, *Management of Wisconsin Soils* by Emmet E. Schulte, Leo M. Walsh, et al, which is an excellent reference for farmers and gardeners no matter where you live.

Nutrient Management Fast Facts, published by the University of Wisconsin's Nutrient and Pest Management (NPM) Program, has all kinds of useful data, including estimates of manure nutrient content. It lists the nitrogen, phosphorus, and potassium nutrient values for average manure from a number of different livestock types and storage systems. It's not perfect, because the numbers in the tables make it seem like all the nutrient value of manure is used up by the end of the second year, which isn't the case. It's still a nice reference. (You can receive this entire publication free of charge by contacting the NPM program at 608-265-2660 or npm@hort.wisc.edu.)

The century-old volume *Farmers of Forty Centuries* by F. H. King made me realize how wasteful we have become as a race with regard to stewarding land, nutrients, and organic matter. Louis Bromfield's classic book *Malabar Farm* helped me to understand the practical realities of renewing tired land with good stewardship practices. *The One Straw Revolution* by Masanobu Fukuoka and Edward H. Faulkner's *Plowman's Folly* challenged my thinking on tillage and weed control in crop rotations. Cornell University cites numerous uses of manure in traditional agriculture around the world, in a summary here: www.tropag-fieldtrip.cornell.edu/tradag/Chapter13Page4.html.

I can't separate the topic of manure from the broader field of soil science. Part of being a student is studying scholarly written works and references, and I have drawn heavily on these to compile this book. These references I've mentioned barely scratch the surface. I encourage you to seek out these and others to broaden and deepen your understanding of both manure and soil management in general. Another part of being a student is studying the world around us — don't forget to stop and smell the roses, along with the manure.

General Formula for Metric Conversion	
Ounces to grams	multiply ounces by 28.35
Pounds to grams	multiply pounds by 453.5
Pounds to kilograms	multiply pounds by 0.45
Cups to liters	multiply cups by 0.24
Inches to centimeters	multiply inches by 2.54
Fahrenheit to Celsius	subtract 32 from Fahrenheit temperature, multiply by 5, then divide by 9

INDEX

Page numbers in *italic* indicate illustrations. Page numbers in **bold** indicate tables.

OTHER STOREY TITLES YOU WILL ENJOY

How to Mulch by Stu Campbell and Jennifer Kujawski
Mulching makes your garden and yard much easier to care for, helping to retain moisture, keep weeds in check, and protect young plants. This concise guide covers how to mulch for any situation, including sheet mulching, feeding mulches, and living mulches for use in the yard, garden, and home landscape.
96 pages. Paper. 978-1-61212-444-5.

The Complete Compost Gardening Guide by Barbara Pleasant and Deborah L. Martin
Following the guidelines of the Six-Way Compost Gardening System, you will learn the principles for improving your garden with healthy compost. This comprehensive tour of materials and innovative techniques helps you turn an average vegetable plot into an incubator of healthy produce and flower beds into rich, blooming tapestries.
320 pages. Paper. ISBN 978-1-58017-702-3. Hardcover. ISBN 978-1-58017-703-0.

The Backyard Homestead Guide to Raising Farm Animals by Gail Damerow
Expert advice on raising healthy, happy, productive farm animals.
360 pages. Paper. ISBN 978-1-60342-969-6.